MORE ADVANCE PRAISE FOR *RAIN*

"Like the weather, there's no predicting the delightful and sometimes disturbing surprises waiting on every page of *Rain*. Whether she's writing about Mesopotamia or the Met Office, Cynthia Barnett illuminates the hidden connections that tie our fate to a precious resource we neglect at our peril."
—**DAN FAGIN,** author of the Pulitzer Prize–winning *Toms River: A Story of Science and Salvation*

"*Rain* is one of the most elegant and absorbing books ever written about the natural world. Writing with grace and imagination, Cynthia Barnett takes you on a journey into the heart of the most elemental force in our lives. An important, revelatory, and thoroughly wondrous book."
—**WILLIAM SOUDER,** author of *On a Farther Shore: The Life and Legacy of Rachel Carson*

"Captivating and compelling, a delightful celebration of precipitation that is brimming with insight. Whether you're desperate for more of it or you just wish it would stop, you'll never think of rain in the same way again."
—**GAVIN PRETOR-PINNEY,** author of *The Cloud Collector's Handbook*

"Barnett's beautifully written book envelops the reader in a warm shower of intriguing history and fascinating science. Anyone who looks longingly at rain clouds, rejoices in a spring downpour, or frets about drought will love *Rain*."
—**DANIEL CHAMOWITZ,** author of *What a Plant Knows*; director, Manna Center for Plant Biosciences, Tel Aviv University

RAIN

RAIN

A NATURAL AND CULTURAL HISTORY

CYNTHIA BARNETT

CROWN PUBLISHERS
NEW YORK

Copyright © 2015 by Cynthia Barnett

All rights reserved.

Published in the United States by Crown Publishers, an imprint of the Crown Publishing Group, a division of Penguin Random House LLC, New York.

www.crownpublishing.com

CROWN and the Crown colophon are registered trademarks of Penguin Random House LLC.

Library of Congress Cataloging-in-Publication Data

Barnett, Cynthia, 1966–

Rain: a natural and cultural history / by Cynthia Barnett.—First edition.

pages cm

Includes bibliographical references and index.

1. Rain and rainfall. 2. Weather. 3. Rainfall anomalies. 4. Droughts. 5. Physical geography. 6. Climatic changes. 7. Earth sciences. I. Title.

QC925.B315 2014

551.57'7—dc23   2014034180

ISBN 978-0-8041-3709-6

Ebook ISBN 978-0-8041-3710-2

Printed in the United States of America

Book design by Gretchen Achilles

Jacket design by Anna Kochman

Jacket photograph: Ryan McVay/Getty Images

10 9 8 7 6 5 4 3 2 1

First Edition

FOR AARON

CONTENTS

IV CAPTURING THE RAIN

V MERCURIAL RAIN

And who art thou? Said I to the soft-falling shower,

Which, strange to tell, gave me an answer, as here translated:

I am the Poem of Earth, said the voice of the rain,

Eternal I rise impalpable out of the land and the bottomless sea,

Upward to heaven, whence, vaguely form'd, altogether changed, and yet
      the same,

I descend to lave the drouths, atomies, dust-layers of the globe,

And all that in them without me were seeds only, latent, unborn;

And forever, by day and night, I give back life to my own origin, and
      make pure and beautify it;

(For song, issuing from its birth-place, after fulfilment, wandering,

Reck'd or unreck'd, duly with love returns.)

WALT WHITMAN
"The Voice of the Rain"
1885

ORIGINS

The rain on Mars was gentle, and welcome. Sometimes, the rain on Mars was blue. One night, rain fell so marvelously upon the fourth planet from the sun that thousands of trees sprouted and grew overnight, breathing oxygen into the air.

When Ray Bradbury gave Mars rain and a livable atmosphere in *The Martian Chronicles*, science fiction purists grumbled that it was completely implausible. In the previous century, astronomers—and writers like H. G. Wells who borrowed from their work to give sci-fi a tantalizing authenticity—had seen Mars as Earthlike, odds-on favorite for life on a planet other than our own. But by the time *The Martian Chronicles* was published in 1950, those odds had changed. Scientists viewed Mars as chokingly dry, impossibly harsh—and far too cold for rain.

Bradbury didn't care to conform to the scientific views of the day. On any planet, he was much more interested in the human story. He created a rain-soaked Venus, too, but not because scientists then considered it a galactic swamp. Bradbury just loved rain. It fit his melancholy like a favorite wool sweater. As a boy, he had loved the summer rains of Illinois, and those that fell during family vacations in Wisconsin. Hawking newspapers on a Los Angeles street corner as a teen, Bradbury never minded a late-afternoon deluge. And in his eighty years of writing every day, raindrops tap-tap-tapped from the typewriter keys into many a short story and every book.

A Bradbury rain could set a gentle scene or a creepy one. It could create moods of gloom, mania, or joy. In his short story "The Long Rain," he made rain a character all its own: "It was a hard rain, a perpetual rain, a sweating and steaming rain; it was a mizzle, a downpour, a fountain, a whipping at the eyes, an undertow at the ankles; it was a rain to drown all rains and the memory of rains."

So often making rain the mise-en-scène for life, Bradbury was onto something. Everyone knows that life could not have developed without water. Life as we define it required a wet and watery planet. But the Earth-as-exceptional-blue-marble story many of us grew up with is, in some ways, as much a product of the human imagination as the warm Mars sea of *The Martian Chronicles*. Modern scientists have good evidence that Earth did not develop as the sole wet and watery orb in our solar system. Earth, Mars, and Venus were born of the same batch of flying fireballs. All three boasted the same remarkable feature: water.

What's exceptional about our blue marble is not that we had water. It's that we held on to it, and that we still do. While the ancient oceans of Venus and Mars vaporized into space, Earth kept its life-giving water.

Luckily for us, the forecast called for rain.

As even-tempered as it grew up to be, Earth started off 4.6 billion years ago as a red-faced and hellish infant. The universe had been unfolding for about 10 billion years. A new star, the sun, had just been born. Its afterbirth—cold gas and dust and heavier minerals and flaming rocks—was flying about, beginning to orbit. The heavy debris gravitated to the sun, where temperatures were well suited for rock and metal to condense. That's why the four planets closest in, known as the terrestrials, are all made of essentially the same stuff.

For its first half-billion years, Earth was a molten inferno some 8,000 degrees Celsius—hotter than today's sun. Scientists call this violent era the Hadean, from the Greek word *Hades,* or hell. Time and again, the young Earth built up a crust, only to see it incinerated by storms of flaming meteors.

Inside the fiery storms, though, was a lining better than silver. Virtually all of the rocks that made Earth had water locked inside. Water is a remarkable shape-shifter, able to change from liquid to solid—or to gas when it needs to make an escape. As meteorites crashed onto Hadean Earth and split apart, they spewed out water in the form of vapor. This was water in its gas form, no different from the steam rising from a boiling pot on the stove. In a sort of geologic burping contest, both the crashing boulders and young Earth's rising volcanoes disgorged water vapor and other gases into the Hadean atmosphere.

All that water vapor would prove an invisible redeemer. Today, at any moment, more water rushes through the atmosphere than flows through all the world's rivers combined. The molecules speed around like pinballs, bouncing off one another, off other types of molecules, off dust and salt from sea spray. Only when air cools do they slow and begin to stick together, latching on to the gritty particles. When billions of them have condensed, they form tiny liquid droplets. Billions of the droplets, in turn, become clouds in the sky. This is the beauty of water vapor: It falls back to Earth as rain.

When Earth was still a molten mass and hot as Hades, the vapor could not condense. Instead, it hissed away into space. Eventually, though, it began to build up in the young planet's atmosphere. Water vapor is a heat-trapping, or greenhouse, gas just like carbon dioxide. The more gas that built up, the hotter and hotter Earth became— melting the forming crust all over again as the flaming boulders continued to fall like bombs from outer space.

About half a billion years after it started, the blitzkrieg began to wind down. As the last of the flaming chunks fell to the surface or hurtled away, the planet finally had a chance to cool. The water vapor could condense.

At long last, it began to rain.

———

I n 1820, John Keats lamented in his narrative poem *Lamia* how little science leaves to the imagination. The scientist's cold philosophy and dull catalog, the English poet wrote, might as well "unweave a rainbow." The mystery of the rainbow—viewed by many throughout the world as a passageway between heaven and Earth—was dashed, Keats charged, when Isaac Newton explained the optical truth: Rainbows are the refraction of sunlight through raindrops.

Yet Newton was anything but unimaginative. To conceive of gravity, he had to imagine Earth pulling an apple from a tree, pulling the very moon into our orbit. It takes such an eye to picture Earth's first rains—the greatest storms of all time. So much of the young planet was destroyed during the Hadean that scientists have scant physical evidence to suggest exactly when the earliest rains began, what they looked like, and how long they poured.*

The best clues to the first rains lie in Western Australia's Jack Hills. Deep in the craggy orange sandstone there, geologists have dug up tiny grains of zircon that clock in as the oldest terrestrial material found on Earth to date. Nature's trustiest timepiece—the radioactive element uranium—dates the tiny zircons back 4.2 billion years. By then, their chemistry suggests, primeval rains had begun to fall and pool on the Earth's crust. Those earliest lagoons likely boiled away repeatedly in the Hadean's grand finale, called the Late Heavy Meteorite Bombardment, which also cratered our moon.

Only when the meteor storms let up could the great rains let loose. By this time, scientists infer, baby Earth was swaddled in vaporous clouds. So many volatiles had built up in the atmosphere, they moiled the sky thicker than a Newfoundland sea fog, blacker than a line of tornadoes on the Great Plains.

———

* Scientists have discovered impressions of the oldest fossilized raindrops in South African rock, their small, rounded indentations a geological braille that tells of a gentle shower falling on hot volcanic ash 2.7 billion years ago. But that's yesterday compared with the Hadean.

Still, Earth's charred surface likely remained so hot that the rains fell only partway to the ground, evaporating again and again. The Stygian clouds grew impossibly heavy. Lightning, its charges tied to the amount of water in the air, illuminated the lonely scene.

The water vapor accumulated in the upper atmosphere for so long that when the surface finally cooled enough for the rains to touch down, they poured in catastrophic torrents for thousands of years. This was the picture the Stanford University geochemistry professor Donald Lowe painted for me when I asked him to imagine Earth's first rains. Lowe is known for his research on the surface of early Earth and the deepest sediments of today's oceans. He grew up in rain-starved California and lives there now, but he spent half his career at Louisiana State University in Baton Rouge, one of the rainiest cities in the United States. And so it is no surprise that he imagines the first rains like the gullywashers of southern Louisiana, so dense that motorists ease over to the side of the road to wait out the deluges that rap on their car roofs like a steel band's drumroll.

I n Ray Bradbury's 1950 story "The Long Rain," later featured in the film *The Illustrated Man* starring Rod Steiger, four Earthmen crash their rocket ship on a Venus drowning in such torrents. The spacemen trek through the sopping Venusian jungle in search of warm, dry shelters known as Sun Domes. They have no gear, not even hats to keep the streaming rain from pounding their heads and seeping into their ears, eyes, noses, and mouths. They have no shelter, for their rocket ship has been contaminated, and Venus's swampy undergrowth drips as much as its skies. As they search for the cozy Sun Domes, the rain drives each man mad.

With his aqueous Venus, Bradbury, this time to the satisfaction of the planetary purists, reflected the common scientific beliefs of the day. The irony is that in the 1960s, real spaceships discovered that Venus was dry as dust. Conventional wisdom swung from the early view that Venus was waterlogged to the hypothesis that it was always parched.

Today, the evidence has most planetary PhDs convinced that Venus once had what Earth had—water vapor condensing to epic rains that turned much of the surface to liquid—but somehow lost it. Mars, too, appears to have begun life with a warm, wet climate, a huge ocean of water covering nearly a third of its sphere, river valleys carved by rainfall, deltas as expansive as the Amazon's.

Like many planetary scientists, David Grinspoon, chair of astrobiology at the Library of Congress, was drawn to the field by the science fiction writers of his boyhood, including Bradbury and Isaac Asimov, whose book *Lucky Starr and the Oceans of Venus* first piqued his interest in the lost Venusian seas. Earth, Mars, and Venus "started out wet," Grinspoon explains, "drenched by the same scattershot rain of planet pieces." Grinspoon is a musician in a funk band called House Band of the Universe. In the case of Mars and Venus, the question is a twist on the old Creedence Clearwater Revival song: What stopped the rain?

Venus, closer to the sun, appears to have become too hot, its oceans vaporizing away. The searing heat kept the vapor from condensing and completing the circle as rain. Remember that water vapor is a greenhouse gas—more potent than carbon dioxide or any other. The more that built up in the Venusian atmosphere with no rain, the hotter the planet became. This cycle, known as a runaway greenhouse effect, usurped the water cycle. Venus cooked.

Mars, on the other hand, became too cold. Scientists believe the red planet was once wrapped in a thick atmosphere that kept it warm enough for bountiful water. From today's dry and dusty Mars, NASA's orbiters and rovers beam home evidence of rain-carved channels, branching rivers, deltas that might have carried ten thousand times the flow of the Mississippi River. Over perhaps hundreds of millions of years, the cozy Martian air turned cold and thin. The rains dried up. The flowing waters vanished.

Water still exists on Mars—frozen in its polar ice caps and rocks, hidden deep below ground, and in a touch of water vapor in the atmosphere. But whatever frigid hydrologic cycle that remains is not driven by rain.

While Venus grew too hot and Mars too cold, Earth kept just the right atmosphere to hold its water in balance—to hold the rains that turned our fiery young planet blue. Those first rains cooled the hell-charred Earth. They filled its craters and crevices until the ground could absorb no more. They spilled across the meteorite-ravaged terrain, cutting channels that became Earth's first rivers. The rains became the land's first lakes. The lakes spread across the steaming landscape like pools in a rising tide. Over years and decades and centuries, the liberated rains filled great basins and became oceans. Over still more time, they seeped down below the land and sea, filling aquifers that now hold far more freshwater than all our lakes and rivers combined.

Somewhere, sometime, the first rains helped lead to the first life. Whether those primordial cells were stirred up in Charles Darwin's "warm little pond" or originated in hydrothermal vents deep in the seafloor as many scientists today hypothesize, the first life required the rain.

Water alone is not enough, Grinspoon explains. Water is "out there," too—in the atmosphere of Venus and in the polar caps of Mars—but it does not sustain a living world on either of those planets. To become our life force, water also had to build up in the skies, move along with the wind, and pour back to the surface, replenishing the waters, lands, and beings again and again.

From those cataclysmic torrents 4 billion years ago to the hydrologic cycle that slakes aquifers, soil, and rivers day after day, rain, as the source of Earth's water, became the wellspring of life. "Sunshine abounds everywhere," the American nature writer John Burroughs wrote in a paean that soaked nine pages of *Scribner's* magazine in 1878, "but only where the rain or dew follows is there life."

Life, and something more. Humans have a natural affinity for rain, grounded in its necessity for civilization and agriculture. Thomas Jefferson constantly watched the sky from his Monticello home in Virginia, where cerulean thunderclouds build along the Blue Ridge Mountains as if matched by Picasso. Jefferson fretted over cloudless days the way that all farmers do. He found relief when storms returned, carrying moisture from the yet-mysterious West. His letters often closed with a word on the rain—or the lack. "Not enough rain to lay the dust," he would lament. Or he'd gratefully share news of "a fine rain," "a divine rain," "plentiful showers."

Sometimes, after writing to his fellow statesman James Madison, who measured rain in a tin cup nailed to the front gate at his Montpelier estate thirty miles northeast, Jefferson would hold off sealing the letter until morning so he could report the overnight showers at Monticello. "The earth has enough," Jefferson concluded after one such update, "but more is wanting for the springs and streams."

Wanting is apropos, a hint at something more. For the story of rain is also a love story—the tale of "certain unquenchable exaltation" that the poet William Carlos Williams felt as he beheld his storied red wheelbarrow

glazed with rain
water.

And for all of history, it has inspired all the excitement, longing, and heartbreak that a good love story entails. The first civilizations rose and fell with the rain, which has helped shape humanity since our earliest ancestors radiated out of Africa when the rainfall tapered off and the forests turned to savanna grasslands. Every culture had its own way of worshipping rain, from Mesoamerican cave paintings exalting rain deities to modern Christian governors who call prayer for a storm.

Rain and two more of its wondrous pride—clouds and rainbows—

have inspired writers, painters, and poets for thousands of years. Homer's *Iliad* is thick with clouds, as is much of the ancients' poetry and prose. The modern poets wrote unforgettably of rain—what Conrad Aiken called the "syllables of water." Other authors awakened in its absence: Mary Austin, Willa Cather, and Wallace Stegner all found their muse in thirsting lands. True, the sun and the wind inspire. But rain has an edge. Who, after all, dreams of dancing in dust? Or kissing in the bright sun?

We long for rain especially when we've gone without. Rain is bliss when topsoil has turned to dust; when springs have vanished; when frogs have gone silent; when fish have rotted to eye socket on dry lake; when corn has blackened on the stalk; when fat cattle have shriveled to bone; when half a billion Texas trees have perished; when bushfires have incinerated Australia; when unthinkable famine has spread through North Africa.

And then, fast as hundred-mile-per-hour winds, a celebration of rain can turn to terror and the deepest grief. Consider the stormy evening of January 31, 1953, in the Netherlands. Dutch families in the coastal provinces of Zeeland and South Holland had gone to sleep in a festive mood. It was Princess Beatrix's fifteenth birthday. Rain, wind, and waves drum-rolling into the jetties had amplified spirits.

By 2 a.m., the apocalyptic North Sea Storm was pushing floodwaters over dikes and wooden barriers "like boiling milk." When it was over, when all the lost souls were finally counted, they numbered 1,835. Half a century later, another biblical flood, Hurricane Katrina in New Orleans, took 1,836.

Rain itself is seldom the deadly factor in a tempest; wind is often the most destructive force in storm disasters. But the North Sea Storm and Hurricane Katrina, like most every flood disaster in history, spun off an eternal human response: People viewed the flooding like an attack of nature, and vowed to fight back. Like every culture before or since, they convinced themselves that humans could ultimately master the rain.

Ancient Rome had its rain god, Jupiter Pluvius. During drought-induced famine, Aztecs sacrificed some of their very young children to the rain god Tlaloc. In Europe during medieval times, when the extreme rains of the Little Ice Age led to crop failures, starvation, cannibalism, and other horrors, religious and secular courts stepped up hunts, trials, and executions of witches, accused of conjuring storms.

America's natives spun in rain dances with tiny bells attached to belts and walking sticks. Their jingling was soft compared with the cannon blasts of the late 1800s, when some settlers were convinced that firing cannonballs, setting enormous fires, or cutting tracts of forest would yield rain. These and other foolhardy schemes helped lure thousands of naïve homesteaders to try to farm some of the driest land in the new nation.

Quackery ultimately gave way to science that birthed the field of cloud-seeding, a chemical milking of clouds. Today, seeding projects continue in the American West and many other parts of the world. The largest efforts take place in China, where government scientists say they coax showers in arid regions by firing silver iodide rockets into the sky.

If cloud-seeding were a real solution, of course, China's Yangtze River would not be drying up, along with nearby lakes, reservoirs, crops, and livelihoods. Parts of the United States would not be in the grip of the most severe drought since the Dust Bowl—Colorado River drying, California's reservoirs dropping, lush croplands turned to dust.

Even as much of the nation suffers drought, other parts endure increasingly extreme rains, and ominous superstorms such as Hurricane Sandy, the largest Atlantic hurricane on record when it slammed into the Eastern Seaboard in October 2012. Yet we carry the presumption of Jupiter Pluvius, strong enough to erect colossal storm barriers to push unwanted floodwater away in times of too much, clever enough to design enormous reservoirs to store precious rainfall in times of too little.

It will turn out that humanity did, in fact, manage to alter the rain. Just not in the ways we intended.

Wrapping our bodies in Gore-Tex and our cities in giant storm gutters, humans crave mastery over the rain. Yet even in the age of precipitation-measuring satellites, Doppler radar, and twenty-four-hour weather streamed to our smartphones, rain does not give up its mysteries. Hundreds of tiny frogs or fish sometimes fall in a rainstorm, as they have since the beginning of recorded history. Despite forecasting supercomputers that crunch more than a million weather-data observations from around the world each day, rain can still surprise the meteorologist and catch the otherwise elegant bride cursing on her wedding day.

We misunderstand the rain at the most basic level—what it looks like. We imagine that a raindrop falls in the same shape as a drop of water hanging from the faucet, with a pointed top and a fat, rounded bottom. That picture is upside down. In fact, raindrops fall from the clouds in the shape of tiny parachutes, their tops rounded because of air pressure from below.

Our largest and most complex human systems often have the rain wrong, too. In the wettest parts of the United States, we construct homes and businesses in floodplains, then lament our misfortune when the floods arrive. In the driest regions, we whisk scant rainfall away from cities desperate for freshwater. Amid the worst drought in California history, the enormous concrete storm gutters of Los Angeles still shunt an estimated 520,000 acre-feet of rainfall to the Pacific Ocean each year—enough to supply water to half a million families.

These paradoxes could not be more urgent today, as we figure out how to adapt to the aberrant rainfall and storm patterns, increasingly severe flooding, and more-extreme droughts wrought by climate change. Globally, the continents recently drew the two heaviest years of rainfall since record-keeping began. Scientists are bewildered by

the controversy over whether human-caused greenhouse gas emissions are to blame for the precipitation extremes. Increased greenhouse
gases push temperatures higher. Higher temperatures cause greater
evaporation—and therefore greater rainfall—where water exists. They
make it hotter and drier where it does not.

Climate change frightens and divides us, to such an extent that
many people simply refuse to talk about it all. But everyone loves to
talk about the rain. Too much and not enough, rain is a conversation
we share. It is an opening to connect—in ways as profound as prayer
and art, practical as economics, or casual as an exchange between
strangers on a stormy day. Rain brings us together in one of the last
untamed encounters with nature that we experience routinely, able to
turn the suburbs and even the city wild. Huddled with our fellow humans under construction scaffolding to escape a deluge, we are bound
in the memory and mystery of exhilarating, confounding, life-giving
rain.

I

—

ELEMENTAL

RAIN

ONE

CLOUDY WITH A CHANCE OF CIVILIZATION

I f you've ever admired a brilliant azure sky, and wondered how it was the heavens that day radiated such clear and dazzling color, you could probably thank a rainstorm. Rain is Earth's great brightener, beginning with the sky. As fine dust, pollution, and other tiny particles build up in the atmosphere, our celestial sphere grows paler and paler, from blue to milky white. A good rain washes the particles away, shining the heavens to their *bleu celeste* best.

On the land, spring rains are the primitive artists, greening hills and valleys and coaxing flowers to vivid bud and bloom. Summer rains are the long-lived masters of color—the steadier they fall on hardwood trees in June, July, and August, the richer reds and yellows ignite the autumn foliage.

Even in the dead of winter, rain's true colors glisten in the rainiest place in the contiguous United States. The Hoh Rain Forest flourishes in a valley between the Pacific Ocean and the Olympic Mountains in the Pacific Northwest. Nearly two hundred inches of rainfall a year create a Tolkienesque landscape of giant trees, moldering logs, and mosses that drape, cover, trail, and hang on every surface like fantastical hair.

In January, the Hoh is brilliant green—Dr. Seuss on an emerald acid trip. Spring-green mosses carpet the forest floor and fallen logs as big as freight trains. Chartreuse licorice ferns beard the tree trunks; matching sword ferns pierce the understory. Darker olive mosses drape their tapestry from boughs, while a greenish-yellow creeper covers the largest branches like evening gloves.

The wet-weather finery debunks rain's gray reputation; its legacy is quite the opposite. Dearth of rain often means dearth of color—dry prairie, dusty sand, desert animals with pale skins to reflect the sun's heat. Many tropical rain forest creatures evolved bright pigments and sharp markings so those of the same kind could find them in the rain-blurred jungle. The vibrancy and patterns of an African butterfly called the squinting bush brown depend entirely on whether it emerges from its pupae in rainy or dry times. The rainy-day butterfly is bigger, brighter, eats more food, and gets more sex.

Rain *is* the sex for the exquisite orchid *Acampe rigida,* a self-pollinator with small yellow petals tiger-striped red. When raindrops splash inside, they flip off the tiny cap that protects the pollen just like an insect would. As the drops hit the teeny catapult of the stipe, they bounce the pollen precisely into the cavity where it must land to consummate fertilization.

Just as rain brightens the natural world, it has colored the human story, too. Our prehistoric ancestors evolved big brains as they figured out how to follow the fitful rain, which continues to shape humanity in profound ways. When George W. Bush beat Al Gore in the 2000 election that hinged on a Florida recount, Democrats blamed flawed ballots. But rain played a history-making role. An exhaustive study of U.S. meteorological and voter-turnout data to test the conventional wisdom that rain on Election Day helps Republicans confirmed that theory and then some: The researchers also concluded that a perfectly dry Florida Election Day in 2000 would have led Gore to win the state, changing the nation's electoral vote and handing him the presidency.

In *Les Misérables,* Victor Hugo mused about how Waterloo, the battle that brought down Napoleon and ended France's dominance as a world superpower, might have been a French victory if not for muddying rains that delayed combat, giving the Prussians time to regroup. "Providence needed only a little rain," he wrote, "an unseasonable cloud crossing the sky sufficed for the overthrow of a world."

We'll soon reflect on the providence and the pestilence, the rain prayer and rain prose. But to appreciate rain's part in the human story, we must first understand how it works. Rain is created with the help of four main forces that, like the letters of the word, have to fall into just the right place. The Olympic Peninsula, home to the Hoh Rain Forest, is an ideal spot to watch these forces—sun, sea, wind, and terrain— come together to make rain.

The Hoh Rain Forest lies along a river of the same name, derived from the Hoh Native American tribe of western Washington. Forest and tribe also share a similarly unfortunate history. Once prospering in vast numbers from the riches of the rain, the Indians and the trees alike were decimated in the nineteenth century, at the end of which the small bands of survivors were consigned by the federal government to protection on a sliver of their former homes: the Indians on a reservation where the river meets the ocean, the forest in a twenty-four-mile stretch of what is now Olympic National Park.

The forest is a four-hour drive northwest of Seattle and its 3.5 million Gore-Tex-cloaked inhabitants. Pass what feels like 3.5 million more log trucks rumbling along Washington's Highway 101; blanch at miles of clear-cutting on national forestland; read large red signs protesting an expansion of the national park ("What's for dinner? Wilderness?"); and finally, take a winding park road east from the coastal highway along the Hoh River, down its glacier-carved valley, and into the surviving rain forest. Once beyond the people and their piles of lumber, it is yet possible to see and hear the true nature of rain.

An anti-noise-pollution organization has designated the Hoh the quietest place in the nation. Far from jets and cars, and with the trail to myself on a winter weekday, I have the sense of hearing rain for the first time. Not the singular applause we are used to, but a symphony of timbres played at softest volume. Drops strike a muffled plunk in the moss, a gentle splat on the muddy trail, a solid thwack against the mammoth logs and tree roots, a quiet pluck on fern fronds, and a louder snap when they hit the maple leaves scattered on the forest floor.

Eye level, the dominating forest features are not the trees, but their torsos—trunks of hulking spruce, firs, hemlocks, and red cedars whose tops vanished into the clouds long ago. The trunk of the world's biggest spruce, called a Sitka, forms crevices tall enough to stand in. Its roots spread with such mass across the forest floor that they have tide pools all their own, trembling with soft drips from the trees and drops from the sky.

My weather app reports a steady rainfall at 43 degrees. The treetops and moss umbrellas must be catching most of it. Only scattered raindrops make it down into the understory. Each one is oddly distinct as it falls through the boughs—slow-moving, easy to see like snowflakes, and illuminated by the shafts of sunlight breaking through the wispy clouds and big-leaf maples.

As children we learn that water is old and constant; we drink the same water the dinosaurs did. But in rain, water feels as new as these drops as they float from the treetops—Earth's essence reborn in every drop.

Here on the Olympic Peninsula, along the lonely northwest Washington coast, the beachcomber's greatest trophy is a glass float from a Japanese fishing net, tumbled ashore after the ride of its life. The collectible orbs carry the vastness of the Pacific Ocean. Connecting East and West, the sea that Herman Melville called "the tide-beating heart of earth" is larger than all the continents combined and carries more than half the planet's water. By virtue of that volume, it is also the planet's master rainmaker.

The Pacific created the Olympic mountain range, which rises to the east of the Hoh. Marine fossils embedded in the snowy peaks tell the story: About 30 million years ago, the tectonic plate carrying the ocean floor collided with the plate beneath the North American continent. As the heavy ocean plate slid beneath the lighter continental one, the seabed crumpled against the land and pushed upward into the Olympic range.

Today, it's as if the Pacific is trying to take back what it lost. On the coastline west of the Hoh, wild surf storms into rock arches and ricochets against driftwood logs stranded on the dark-pebble beach. Isolated rock formations called sea stacks stand inshore as if at guard, grown with shorn trees that reveal the strength of the wind here. The massive logs, piled like pickup sticks, testify to the power of the sea.

In a single frame, the dramatic panorama captures the building blocks of rain: sun and ocean, wind and terrain.

The sun's energy pulls vast amounts of moisture from Earth's surface, evaporating liquid water into our essential gas, water vapor. The ocean—and it's really a single ocean, despite our divvying it up into the ancient seven seas or today's five—accounts for the bulk of Earth's moisture. Its job is to carry all that water until it is time to hand off to the wind.

Global rain patterns generally follow those of the prevailing winds, such as the trade winds or the fast-flowing currents of air known as jet streams. Rain is most abundant where air rises, and skimpiest where it sinks. In the tropics, the trade winds converge and the heat pushes air upward, building the dark-castle clouds known as cumulonimbus, heavy with rain. So regions around the planet's equator-middle are typically wet. As air moves away from the equator it cools and sinks, creating two dry bands around the globe at the subtropics. These are home to many of the great deserts, from North Africa's Sahara to America's Mojave. Meteorologists commonly describe the subtropical climates as a belt, not a particularly helpful image because they wrap around the planet in two belts, one above and one below the equator. I like to think of the subtropics as Mother Earth's bikini.

Here in the Pacific Northwest, the winds generally travel from west to east across the ocean, blowing strongest along a jet stream a few hundred miles across and a few miles deep. As they speed thousands of miles from Japan over just a few days, the winds scoop up the evaporating water and sun's warmth from the sea. After this warm, wet air roars past the sea stacks and the pebbled beach, past the driftwood logs and coastal cliffs, it runs smack into the Olympic range.

When air meets a mountain, it has nowhere to go but up. Remember rain's MO: abundant where air rises, lightest where air sinks. So the world's rainiest places are usually found on the windward side of mountains, looking to the sea. The most extreme rain in the United States occurs in the North Pacific on Hawaii's emerald-green Mount Waialeale on the island of Kauai, soaked with 460 inches a year. Slightly higher than that, the worldwide rainfall records are set in the northeast corner of India in the state of Meghalaya overlooking the Bay of Bengal. In 1860, a village there called Cherrapunji logged the greatest rainfall in recorded history—1,042 inches in one year. Every rain-obsessed scientist I interviewed seemed to dream of visiting there.

On the Pacific side of the Olympics, the moisture-laden air drizzles between 140 and 170 inches of rainfall a year onto the Hoh Rain Forest. As the spent clouds continue over to the east side of the range and descend into the lowlands, rainfall decreases as well—creating what's known as a rain shadow. This is why cacti thrive in the town of Sequim, Washington, east of the range and less than thirty miles from white-crowned Mount Olympus. A fast-growing haven for Puget Sound retirees, Sequim sees only 15 inches of rainfall a year—typical of Southern California.

The mountains likewise keep Seattle from having much rainfall. You read correctly. Notwithstanding the deluges and lightning that pound outside Frasier Crane's apartment on the TV sitcom *Frasier*, Seattle is not one of America's rainier or more lightning-struck cities.

Rather than flashy thunderstorms, Seattle is known for salmon-silver clouds that drizzle through winter, just the way I found the skies on the morning I drove to the University of Washington to interview the weather pasha of the Pacific Northwest, atmospheric sciences professor Cliff Mass. I hoped Mass could help me clear up some common misconceptions about rain. I live in Florida in the waterlogged South, the rainiest region in the United States. Seattle gets little rain in comparison. In the global swirl of the atmosphere, what makes the rain so different on the same continent? And why does Seattle, nicknamed "City of Rain," have such a soggy reputation?

A native New Yorker, Mass was drawn to meteorology as an undergraduate at Cornell, where he worked in Carl Sagan's planetary studies lab. He published his first academic paper—a model of the Martian atmosphere—with Sagan. They found the circulation of weather on Mars "strikingly tied to topography." So, too, is Seattle's weather.

Mass, who has a popular book, blog (cliffmass.blogspot.com/), and public radio program on the weather of the Pacific Northwest, spends a lot of his time debunking his adopted city's rain refrain. To its west, Seattle is sheltered by the Olympics. To the east, the rugged Cascades stretch in a startling line of volcanoes from British Columbia all the way to California. As air crosses the coastal mountains and descends into the lowlands of western Washington and Oregon, annual rainfall decreases, from more than 100 inches a year to between 25 and 45 inches. This rain shadow covers an urban corridor from the Canadian border all the way to Eugene, Oregon—making Seattle and Portland drier than any major city on the U.S. Eastern Seaboard, with 20 inches less rainfall than Miami, 5 less than New York and Boston.

Mass pulls up the National Weather Service's precipitation map—rainfall pulsing in electric blue—to show me how rain moves across the continent. During winter, the jet stream sends decent rainfall to both sides of North America; the Pacific Northwest gets more than half its rain between November and February. But come summer, the Pacific jet stream weakens. The Pacific Northwest gets hardly any rain. As Melville's sea stays cool, the Atlantic Ocean and the Gulf of Mexico grow warmer and warmer. The eastern half of the country endures lightning strikes, washed-out picnics, tropical storms, and hurricanes through summer, while Professor Mass hosts a "Dry Sky Party" in Seattle the last week of July. Almost always a drought week, it is also the answer to his most-frequent question from the public: "What date should we choose for our wedding?"

Mass can put grooms and brides at ease about the rain, but makes no promises of azure skies. The sea still sends its clouds during summer. But, idling low between the mountain ranges, they don't do much more than block the sun. The real reason for Seattle's rainy character is the cloud cover. Mass explains that Seattle and Portland are blanketed by clouds about 230 days a year, compared with 160 in Boston and 120 in sunny Miami. The number of days with just a trace of rain is also considerably greater over the Northwest than back east.

While Seattle experiences thunderstorms only about 7 days a year, the U.S. record-holder, Lakeland, Florida, watches thunderstorms roll in over its slate-colored lakes about 100 days. Lake Victoria in Uganda, Africa, is the world's thunderstorm capital, rumbling an average of 242 days a year.

Thunderstorms need convection—warm, moist air currents that rise quickly to form the dark-tower cumulonimbus. This is the same convection we know from cooking; the little eruptions caused by rising heat in a pot of, say, seafood gumbo simmering on the stove. The Gulf of Mexico cooks up the most memorable of both.

The ten rainiest metropolitan areas in the United States are all in the Southeast, most doused by storms brewed in the warm waters of the Gulf. Its surface temperatures hover in the 80s. Tropical storm season, between June and November, spins off far more rain than Seattle sees all year. Gulf-front Mobile, Alabama, is drenched with more rain than any other metro in the nation, 65 inches a year. New Orleans and West Palm Beach come in second and third, both just over 62 inches, followed by Miami and Pensacola. As it moves up and out of the Southeast, this summertime air, heavy with moisture and warmed by the Gulf and Atlantic, continues across the eastern half of the continent, where it is further heated by the sun. This turns on the convection, roiling up thunderstorms across the eastern United States. The storms flash summer rains and lightning from the Atlantic Seaboard all the way to the Rockies. That's where the action slows. Air meets the mountains and rises. The ascending clouds release their rain, this time on the east side—creating a rain shadow at the gateway to the arid West.

To the west and south of the Rockies lie America's driest places, which lose more water to evaporation and transpiration from plants than they receive from the skies. (Transpiration is the way moisture moves through plants, from the roots up to small pores on the underside of leaves, where it changes back to water vapor. An acre of corn can send 4,000 gallons of water back into the atmosphere each day.)

Those suffering from ombrophobia—fear of rain—might consider settling in Yuma, Arizona. It seems unfair to label Yuma the driest city, straddling as it does a bucolic little bend of the Colorado River. But Yuma is the rain-scarcest city in the nation, averaging a three-inch trickle in a year. Not infrequently in the wide skies over Yuma and other parts of the arid Southwest, residents watch sheets of rain begin to unfurl from auspicious purple storm clouds, backlit by the sun. But the rain stops halfway, hanging mid-horizon like a magician's trick. Known as rain streamers or by their scientific name *virga,* the half-sheets evaporate into the dry air before the rain can reach the ground. "Torture by tantalizing, hope without fulfillment," Edward Abbey called the withering curtains in *Desert Solitaire.*

The world's dry places have their seasons of rain, often called mon-soons, from *mausim*, the Arabic word for season. Come summer in the Southwest, North American monsoon clouds dollop white peaks atop the low-lying desert mountains, signaling the rains to come. When the showers finally kiss the parched desert land, they release a perfume of sagebrush and creosote befitting the special occasion. We may be losing our regional dishes and dialects, but rain is always vernacular—preserving indigenous scents, language, and customs wherever it travels.

Many of the East Coast transplants who move west for the desert sun don't realize how much they will miss the rain that was so an-noying when it washed out Fourth of July fireworks back home. In droughty recent years, rain's arrival roused applause in the streets of Los Angeles, sent Albuquerque residents into front yards to celebrate, and lit up social media across the West with squall-gray selfies, rain hashtags, and umbrella exclamation points.

Of all the rains that soak Earth, none are more celebrated than those that blow in with the Asian monsoon. Scientists define the monsoon as a great wind; it's essentially Earth's biggest sea breeze. But to most of humanity, the monsoon is a miracle of seasonal rain. Almost two-thirds of the world's people live by the rhythms of the spectacular inundations that bring water for drinking, farming, and life's other fundamentals. The annual drama makes monsoon season something of a holiday for people across south and east Asia, who throw colorful festivals, sing monsoon ragas, go for monsoon cures, hook up for monsoon romances (the Eastern version of the summer romance), and dance barefoot in deep street puddles to rejoice in its arrival.

The monsoons can also bring peril; floodwaters can kill hundreds, sometimes thousands of people in China, India, Nepal, and surround-ing regions, displacing millions of others. But few disasters have been worse than the intermittent failure of the Asian monsoon, source of some of history's most grievous famines. Even today, because the crops and water supply of entire nations, including India, depend upon its strength, failure of the monsoon can crash markets, spike food prices, provoke suicides, trigger energy shortages, and swing national elections.

Monsoon meteorology boils down to temperature differences between land and sea. Like other rains, monsoons are driven primarily by the sun, constantly drawing up all that moisture from the oceans. As summer approaches, old Helios beats directly down on the countries bordering the Indian Ocean and the South China Sea, making the lands much warmer than the waters. As air heats over the land, it becomes lighter and rises. This causes cooler, wet air originating at sea to change direction and rush to the land. When these fast and fluid atmospheric freight trains meet the highest mountains in the world, the Himalayas, they dump load after load of rain. This is how the villages of India's state of Meghalaya—the name means "abode of clouds"— balanced on the steep Khasi Hills overlooking the Bay of Bengal, are drenched with the greatest rainfall in the world. I pined to see those hillsides during monsoon season.

Monsoons are influenced by such a dynamic mix of sea, land, and atmospheric conditions that computer models are not yet good at predicting a strong, weak, or failed season. Even rain's most predictable patterns can be knocked off kilter by what scientists call teleconnections—faraway shifts in the atmosphere that reverberate halfway around the world. The best-known is El Niño, an unusually warm sea-surface temperature in the middle of the tropical Pacific Ocean. Every three to seven years, El Niño arrives to massively shake up rainfall on every continent. It can weaken the Asian monsoon, bring torrential rains to the western United States and scorching drought to Australia, and tamp down hurricanes in my part of the world.

Modern living, with its reservoirs, irrigated agriculture, and invisible water pipes that run for millions of miles underground, dulls us to our fragile dependence on rain. El Niño—named for the Christ Child in the 1600s by South American fishermen who first observed the warming that tends to come around Christmastime—is one dramatic reminder. Human-caused climate change is proving another. Scientists say a warming globe will double extreme El Niño events, the sort that send California mansions sliding into the sea and set bushfires raging in the Australian outback. Climate change is also causing an alarming worldwide spike in torrential rains. If there is one point of solace, it is that we have lived through them before.

Scientists and historians are careful not to claim that climate was the dominant force in developments such as agriculture and the rise and fall of civilizations. But in the past few decades, proxy records reconstructed from tree rings, tiny pollen grains from ancient swamps, and cores drilled from glaciers, lake beds, and the ocean floors have revealed profound connections among human progress, aridity, and rain. The anthropologist Brian Fagan calls climate "a powerful catalyst in human history, a pebble cast in a pond whose ripples triggered all manner of economic, political, and social changes."

In this case, the ripples fan out from a drop of rain. The same one ruining a seaside wedding today could have fallen millions of years ago on our primate Adams and Eves.

In the course of geological time, rain can fall heavily for hundreds or thousands of years. Scientists call such a wet period a *pluvial,* from the Latin word for rain, *pluvia.* Before humans split off from their primate cousins, our earliest ancestors lived in pluvial times. To travel through the storms of history and imagine prehumans and their lives in the rain forests of East Africa, take a look at your fingers after a long bath or an afternoon spent swimming.

We've all seen our fingers and toes shrivel up like alien extremities after a good soak. Popular belief long held that this prune effect is caused by osmosis—our digits must absorb water and swell up, rippling the skin into little folds. But a neurobiologist named Mark Changizi had another idea.

In 2008, Changizi was researching the shape of human hands when he came across a surgical paper from the 1930s documenting that patients with damage to their arm nerves don't get wet wrinkles on their fingers. It turns out that doctors who work with nerve-damaged patients already knew pruniness couldn't be an accidental side effect of wetness. Instead, our water wrinkles are the work of the autonomic nervous system—a largely involuntary control panel in the lower brainstem that also prompts basics such as breathing, swallowing, and sexual arousal.

Changizi put aside his question on the shape of hands for another within grasp. Other primates, such as macaques, get the finger and toe wrinkles, too. An automatic physical trigger to water suggested to Changizi that they must be some sort of adaptation. Primates would not have evolved to deal with porcelain bathtubs or Olympic-sized swimming pools. But they would have had to adapt to rain.

Changizi had a new grad student, Romann Weber, working in his Boise, Idaho, lab and asked him: "Can you think of a good reason to get pruny in the rain?" Weber gave it some thought. "Rain treads?"

In dry conditions, smooth tires like those on race cars provide the best grip. On wet roads, though, rain treads handle much better. Changizi and Weber hypothesized that smooth fingertips likewise give humans the best grip in dry times, but wrinkly ones might help us hang on when it's wet. Their subsequent research indicates that's true and, moreover, that human evolution appears to have gone a step further to best the tire.

Magnified, our finger wrinkles look like drainage channels: rain chiseling a landscape for thousands of years. They are not perfectly circular, but flow downward like a river branching into an estuary. The channels move farther apart as they flow down, like mountain topography shaped by streams flowing toward the sea. The channels are also pliable—so pressing your finger on a surface squeezes water out through them.

The wrinkles appear nowhere else on our bodies but the hands and feet. They form quickly enough to gear up for rainy conditions, but not so fast that casual contact with water, such as eating fruit, shrivels us up. All in all, Changizi believes, these are clues to an ancient adaptation for gripping in the rainy forests where human ancestors lived some 10 million years ago.

On the family tree of human evolution, the branch between the fruit-noshing primates of East Africa and the chai-drinking moderns of today is represented by the *hominids*. Since Charles Darwin's time, science textbooks have attributed their emergence to the "savanna hypothesis" of evolution: For millions of years, our primate ancestors stuck around their Eden-like forests. Why would they need to stand up and stretch out when plenty of fruits, berries, and other tasty nourishment hung so easily within reach? It was not until the rains tapered off, and the forests with them, that our primal foremothers and fathers went on the move. Their rain forest home gradually gave way to wide, open savannas. A new sort of primate began to walk on hind legs, and developed a long field of vision to look out over the grasslands.

New fossil discoveries and advances in DNA research are proving our evolution from tree to tea was a lot more complicated. Emerging paleoclimate and fossil evidence still supports the theory that hominids developed their ability to walk on two legs as the region's rains tapered and forests transitioned to savanna grasslands between 5 and 7 million years ago. But anthropologists now see that human evolution was not just survival of the fittest, but survival of the most adaptable to changing climate. The major leaps in our evolutionary history line up with dramatic changes in prehistoric climate, namely temperature and rainfall.

As arid periods dried Africa's rains, cracked the earth beneath our ancestors' feet, and evaporated lakes and rivers, many of the animals and plants that had thrived in the waters and wet forests went extinct. Digging in an East African lake bed in 2012, the American paleontologist Christopher Brochu discovered the skulls and jaws of an ancient crocodile that grew up to twenty-five feet long. Brochu says the terrifying beast, the largest predator in its environment, probably dined on the early humans who came to the lake.

But on the evolutionary menu, the adaptable hominids on the way to becoming human ate the crocodile's lunch. They outlived this predator and others, moving from place to place during drought and developing the first stone tools to help them cope with the new types of predators that took the beastly croc's place—becoming the hunter instead of the prey.

The most recent 2.5 million years have been a seesaw of wide climate swings between epic rains that spread monstrous lakes across Africa and severe aridity that sucked those lakes back to dust. With each cycle of wet and dry, new species came and went. Yet ultimately, *Homo sapiens* persisted, even when their numbers plummeted to several thousand hardy souls in the wake of a series of mega-droughts. Many archaeologists believe *Homo sapiens* built their big brain power during these rain-starved times, evolving speech to share what they knew about water and food to survive famine. Others argue the periods of huge lakes and high moisture will prove more important in the story of human evolution. But whether humans rose to adapt to drowning pluvials, dusty aridity, or neither, the scientists agree that a large part of what makes us human is our remarkable ability to adapt to any conditions the atmosphere blows our way, at least any seen on Earth in the past few million years. It is the same trait that allows us to live on all seven continents, from bamboo houses on the rainiest hillsides in the world in India's state of Meghalaya to faux adobe minimansions in the Mojave Desert.

Around 30,000 years ago, in another extreme climate swing during the last big Ice Age (appropriately named the Last Glacial Maximum), our evolutionary cousins the Neanderthals went extinct—even though they seemed ideally built for cold. *Homo sapiens,* their bodies seemingly made for the rain-fed tropics, continued to expand into icy Europe and around the world. After millions of years of evolution through the pluvials and the mega-droughts, we became the sole survivor of all hominids, and among the most adaptable species in Earth's history.

O ur Ice Age images are frozen in massive ice sheets and glaciers and packed snow (and a dopey woolly mammoth that talks—in Ray Romano's Italian New York accent, no less). But this difficult stretch of early human history was also distinct for its dryness. Despite the flood tragedies and the crop-killing torrents of history, survival proves much tougher without rain than in times of too much.

With so much of the planet's water locked up in ice, Earth's great evaporative engine had little fuel for rain. Sea levels dropped by more than three hundred feet. In arid sweeps of northern Africa and southern Australia, rainfall tapered to less than half its previous bounty and today's. Their rivers and lakes vanished—along with sources of food. In a loop of depletion, less rain meant fewer trees to release water back into the atmosphere, leading to yet less rain. Most of the world's deserts expanded. Harsh winds mounded up great sand dunes and pushed them around like Brobdingnagian chess pieces. Plants and trees could no more survive here in the desiccated soils than they could in the ice sheets draped over the Northern Hemisphere.*

Remember rain's role in scrubbing the sky blue? Lack of rain means lots of dust; our Ice Age ancestors gazed up at a bleak celestial sphere. Studying the sediment cores of the Last Glacial Maximum is

---

* While most of North America was covered in ice, there were regional exceptions to the aridity. Winds swooped glacial air to warmer regions including today's western United States, bringing a record pluvial that created mega-lakes. Lake Bonneville spread across almost all of Utah before it turned into its famous salt flats.

like peeking under the bed: Scientists find dust everywhere they look. Today, the North African and Arabian deserts swirl with the greatest dust storms on Earth, mountainous yellow-brown clouds known as *haboobs*, visible from space. Cores from the Arabian Sea suggest the Ice Age skies were 60 percent sandier.

Early modern humans endured the cold, dust-dry times as hunter-gatherers in small, isolated groups, having spread from Africa into Europe and Asia and down into Australia. As Asian Ice Age hunters tracked caribou and musk oxen across the inhospitable white tundra we know as Siberia, they came upon a land bridge exposed by the low seas, and now humanity began to trickle into the Americas. But across all the inhabited continents, humans didn't really spread out and settle down until another dramatic swing in climate—this time, to warmth, and the steadier rains of our own epoch.

Planetary scientists suspect changes in Earth's tilt and orbit helped spur the great freezing and in turn the great melts, including the one about 18,000 years ago that began to release water back into our familiar cycle of rain, toward the mild climate modern humans have generally enjoyed since.*

A drumroll is in order for the Holocene. After millions of years of heave and ho between dramatic cooling and warming, parched times and pluvials, humans in the most recent 12,000 years have lived in an unprecedented stretch of relatively stable climate. "Relatively" is the clincher. All human civilizations arose during this climate halcyon that the anthropologist Fagan calls the Long Summer. But century-scale extremes—epic droughts and ceaseless rains—regularly threaten the calm, like hurricanes slamming a normally sunny beach hideaway.

---

* Paleolithic people did endure another cold and rainless millennial 13,000 years ago, the Younger Dryas. The name comes from an alpine flower, white dryas, whose ancient pollen in ice cores shows scientists when tundra replaced forest in the Northern Hemisphere. The Younger Dryas may have been sparked by an asteroid impact or mega-flood of North American glacial meltwater into the Atlantic that disrupted the ocean's circulation for a thousand years. It appears to have slowed the march of humanity in the Northern Hemisphere; populations declined or plateaued in all lands north of the equator.

Humans began the Holocene as happy hunters, soon to be home owners. Increased rainfall and warmer temperatures greened North America and Eurasia, and expanded the amount of livable land. Likewise in Asia, as the North Atlantic basin warmed, summer monsoons strengthened, spreading a tropical green garden of trees and plants across the region.

The kinder climate brought population growth and more permanence; a trade up from caves and pits to huts made of wood or stone. Radiocarbon dating lets scientists analyze tiny bits of seed and bone; this is how they know that people in many parts of the world lived well on wild abundance, filling up on game and native plants, fruits and nuts. This is also how they track a decline, about 8,000 years ago, in greenery across the land. The new wave of aridity coincides with a shift from hunting wild sheep to herding them, gathering wild grains to growing them.

Why the hunter-gatherers set aside spears and harpoons for sickle blades and grinding stones has been one of the great puzzles of archaeology. But the dawn of agriculture does line up with a worldwide trend of miserly rainfall, including a weakening Asian monsoon. Many scientists believe the faltering rains and worsening aridity clustered people along rivers, bringing smaller communities together to build irrigation systems and feed themselves and their neighbors. The four early civilizations of the ancient world arose in great river valleys: the Nile in ancient Egypt, the Tigris-Euphrates in Mesopotamia, the Yellow River in China, and the Indus in ancient India. Agriculture was our way of adapting to stretches of drought. For these four civilizations, it would not be enough.

In the same way Earth can support life having swung into what astrobiologists call the Goldilocks zone between Mars and Venus, rain has its just-right sweet spot for humans. Rain's temperament can mean the difference between food and famine, health and plague, social unrest and national content. Years of too much can bring plague and pestilence. Years with too little stretch on with hunger and desperation. In the end, no condition proves as devastating as when rain disappears—extreme drought that can last hundreds of years. As miserable as it can sometimes be, humans have never been able to live without their color-giving rain.

DROUGHT, DELUGE, AND DEVILRY

n northwest India and Pakistan, the largest of the early civilizations in the ancient world lies buried under golden sand. Unlike the Egyptian tombs or the ziggurats of Mesopotamia, most of the lost cities of the Indus Valley Harappan people are not yet excavated, their mysterious script not fully deciphered. But archaeologists working in the ancient cities of Harappa and Mohenjo Daro have revealed a remarkably advanced culture beneath the desert.

Nearly 5,000 years ago, the Harappan people lived in planned cities with walls, wide streets, reservoirs, and the first urban sanitation works. Their brick homes had hearths, wells, pipes, and rooms for bathing. Granaries and other agricultural ruins hint at major farming and trade networks that served tens of thousands of people.

All of this life flourished on monsoon rains and rivers. The vast majority of Harappan sites have been found along large river channels carved into the desert, now ghost rivers visible only in satellite imagery. Around 4,000 years ago, the people began to abandon their advanced cities. Harappan civilization slowly disintegrated.

Ever since the lost cities of the Indus Valley were discovered in the 1920s, scholars have disagreed about the reasons for their demise. The Hindu holy text known as the Rig Veda, written in an early form of Sanskrit beginning about 1750 B.C., held an intriguing clue. The Rig Veda describes a mythical river that once ran parallel to the Indus

Valley, flowing from the Himalayas all the way to the Arabian Sea. Appearing in the text more than seventy times, the Saraswati River was often interpreted as metaphor. But scientists say that it was something more.

In recent years, research into the region's past climate has uncovered a dramatic shift in rainfall patterns. Searching in an ancient rain-fed lake in northern India, paleoclimatologists using radiocarbon dating have discovered that 4,100 years ago, the summer monsoons began a rapid decline. They did not return to normal for two centuries.

For an unimaginable two hundred years, the Harappan region saw hardly any rain. Around the same time in China, Egypt, and Mesopotamia, the three other earliest-known civilizations also were lost to the dry sands of history.

I n the 1920s and '30s, the British anthropologist C. Leonard Woolley— think Indiana Jones with high cheekbones and kneesocks—led the excavation of Ur, one of the largest and wealthiest city-states in ancient Mesopotamia (from the ancient Greek for "land between the rivers") at the confluence of the Tigris and Euphrates in the Middle East. Newspapers worldwide reported on the tombs he uncovered in Ur's royal cemetery, and the extravagant treasures inside: jewelry and garments made of gold, lapis lazuli, and carnelian; bowls of alabaster; solid-gold weapons; bejeweled animal figures mythical and real. The public was captivated particularly by the tales of small cups near the remains of attendants who died with their masters, headline-making evidence of ritualistic suicides. One tomb held seventy-four of the sacrificial victims.

The book of Genesis names Ur as the birthplace of Abraham. The son of a clergyman, Woolley tapped his biblical knowledge and the passion of his faith to draw people to his finds. Tourists including European royals and the queen of mystery, Agatha Christie, descended on the remote Iraqi desert from around the world to see its excavation. The crime author took the Orient Express to Baghdad and hung around Woolley's dig to gather color for her book *Murder in Mesopotamia*.*

Woolley was convinced that he would next find proof of Noah's flood. He remained so focused on the deluge that he may have set archaeology back half a century in finding the real event—quite the opposite calamity—that devastated Mesopotamian culture.

Along with their fiction-inspiring burials, Mesopotamian people cultivated Western civilization well before the ancient Greeks. The Sumerians, who built a dozen cities including Ur, carved into clay tablets with wedge-shaped styli to leave the first record of written language. They conceived the first written philosophy and laws; in architecture, the column and the arch; in math, the concept of zero. But a little over 4,000 years ago, three-fourths of Mesopotamian settlements were abandoned during an epic drought that seared for three hundred years. Some of the civilization's greatest contributions, including the written and spoken languages, were lost for thousands more.

If only for some rain, students of the classics might be practicing their ancient Sumerian, alongside Latin and Greek.

About 4,300 years ago, a king called Sargon of Akkad seized power in Mesopotamia, grabbing both the agricultural hinterlands in the north and the growing city-states to the south. Over the next century, Akkad and his sons and grandson built the world's first empire, ruling from the headwaters of the Tigris and Euphrates all the way to the Persian Gulf. The rain-fed northern plains fueled a large and lucrative agricultural bureaucracy. Their written records show the Akkadians controlled and taxed production and traded long distance for barley and wheat.

---

* She modeled the book's victim, killed with a large blunt object to the head, on Woolley's wife, Katharine, who was widely disliked but so capable that she led the final season's dig at Ur. Katharine Woolley also may have been a matchmaker: Christie ended up marrying the young archaeologist-in-training Mrs. Woolley enlisted to show her around.

In what is now northern Syria, the Akkadians erected monumental buildings at Tell Leilan, with city walls, expert canals, huge agricultural projects, and an impressive military. And then, after a hundred years of prosperity, people across the empire suddenly abandoned the cities and farms.

Carved in cuneiform script in the wake of the fall, an epic poem called the Curse of Agade (or Akkad) blames divine retribution from Enlil, a raging storm god responsible for the worst of floods and droughts. The composition describes "great agricultural tracts" that "produced no grain" and "gathered clouds" that "did not rain." Scholars always read the curse as metaphor, much like the lost Saraswati River of Hindu lore. They blamed the collapse of the empire on managerial incompetence, overpopulation, invasion, and various other theories.

But in 1993, the Yale archaeologist Harvey Weiss, who heads the university's Tell Leilan excavation, essentially declared that the cuneiform scribes had it right. The city once so grand that its walls still rise above the Khabur Plains had been blanketed with windblown sand for three hundred years, a time of virtually no human habitation.

In Tell Leilan and beyond, at site after site across the Habur and Assyrian Plains, Weiss and his colleagues find cities abandoned 4,200 years ago—no grains, no ceramics, no signs of humanity for three centuries, just a thick veil of dust. With no rain to moisten the soil for all that time, even the earthworm holes disappear.

The desiccation spread far beyond the Middle East. Lake-bed soils in Africa, dust content in ice cores from northern Peru, ancient stalactites hanging from caves in China and India, and pollen samples from North America all point to a worldwide crisis of drought, with aridity off the charts compared with the rest of the Holocene. Conditions looked more like the Younger Dryas, which decreased human populations across the Northern Hemisphere 13,000 years ago.

The rainless centuries line up with not only the downfall of Mesopotamia and the disappearance of the mighty Harappan civilization, but also the collapse of the Old Kingdom of Egypt along the Nile.

In China, scientists note the demise of a number of Neolithic populations, a shift from agricultural-based cultures back to pastoralism, and a marked decline in the number of archaeological finds along the lower basins of the Yangtze and Yellow rivers.

Time and again through the Holocene, as rain goes, so goes civilization. For more than ten centuries until A.D. 900, the Maya flourished in the lowlands of Central America, reaching a population near ten million and reliant on delicate water management in a drought-prone terrain. Like the great civilizations of the Indus, Tigris-Euphrates, Nile, and Yellow rivers, the Maya could overcome droughts that stretched years or even decades. But a three-hundred-year dearth of rain—lake-bed cores show that it lasted from 750 to 1025—proved too much.

If drought brings death in slow motion, we think of deluge as wielding it in a flash flood. In most cases, rain is not the deadly element in flood; that is wind and tidal surge, the forces that washed away up to half a million souls in what is believed the deadliest storm in history, the Great Bhola Cyclone of 1970 that drowned Bangladesh at the Bay of Bengal. Hurricane Mitch in 1998 was one great exception to the rain rule. Its fierce winds, waves up to 44 feet, and astonishing rains destroyed vast regions of Central America. The mountains of Honduras and Nicaragua milked as much as 75 inches of rain from the hurricane, which created great floods and mudslides that swept away entire villages and their inhabitants. More than 11,000 people were killed and 3 million more left homeless.

Like the worst droughts, though, the worst deluges have often been those that unfolded slowly, dragging on for decades or centuries. For all its color and life, too much rain for too long settles with grim darkness: mold, rot, and floodwaters that never seem to drain; clouds of mosquitoes outside and fevered disease within; dread to inspire Shakespeare's *Macbeth*.

This may never have been truer than during the Middle Ages, when Europe wasted in some of the heaviest rains in human history. In the fifteenth-century woodcut *The Four Horsemen of the Apocalypse* by the German artist Albrecht Dürer, foreboding storm clouds frame the four riders—Death, Famine, War, and Plague/Pestilence. Endless rains could presage at least three of them.

In medieval Europe, the 1300s marked the beginning of a five-century climate shift known as the Little Ice Age. Persistently copious rains, floods, snows, and early and late frosts led to widespread crop failures, famines, social instability—and fear that would ignite paranoia. The second decade of the 1300s was the rainiest in a thousand years. Severe, near-constant storms thundered down in brutal surprise. One wet summer followed another. Temperatures were exceptionally cold. Electrical storms brought steady and severe lightning strikes. People rarely saw the sun, whose absence ruined grape harvests and salt production. Chroniclers of the day often used the metaphor of the Great Flood to describe the conditions; they believed God must be punishing humans for their sins, including the wars under way around Europe.

In 1315, the downpours began on Pentecost in May and continued "almost unceasingly throughout summer and autumn." Floodwaters ran so deep through Malmesbury, the oldest borough in England, its chronicler "thought that the prophecies in the fifth chapter of Isaias were being fulfilled." Throughout northern Europe, the deluges broke through mills, bridges, and other businesses and infrastructure. The Danube burst its banks three times in Austria and Bavaria. On Austria's Mur River alone, high waters swept away fourteen bridges.

Along the coasts, storms and floods battered Normandy, and Flanders at the North Sea. That August, King Louis X of France planned a military attack on Flanders. But rains soaked the advancing soldiers day and night, "in most miraculous fashion such as no mortal then living had ever seen." The lowlands of Flanders turned to bogs. When the soldiers galloped forth, "so wet was the ground that the horses sank into it up to their saddle girths," wrote the medieval historian Henry Lucas. "The men stood knee-deep in the mud, and the wagons could be drawn only with the greatest difficulty." Stuck in their flooded tents and running short of food, the French soldiers retreated. The Flemings thanked God for the rains. At least until famine set in.

From the Pyrenees mountain range in southern France to England and Scotland, across the Holy Roman Empire and east into Poland, the torrential summer rains prevented grain from maturing. That meant no autumn seeding of wheat and rye. Then spring rains made it impossible to sow oats, barley, and spelt. Hay could not be cured. In England, the price of wheat quadrupled in 1315, and then doubled again by the end of the year.

In the prior century, population growth had spread many rural families onto marginal farmlands that could eke out crops in good rains. The lands proved inadequate for absorbing downpours that cut deep gullies into the countryside and washed off thousands of acres of thin topsoil. In some rural areas, as much as half the arable land vanished. Crops failed. Herds and flocks withered. As domestic animals and fowl became scarce, so did meat and eggs. The anemic grain had little nourishing power. Families foraged in the sopping fields, gnawing on leaves and roots.

Of the many famines that have struck since the rise of agriculture, the Great Famine of 1315–1322 is one of the few ascribed largely to unremitting rains. Without harvest, people became so hungry they turned to eating dogs, cats, and horses. "Horse meat was precious; plump dogs were stolen," wrote the English Benedictine Johannes of Trowkelowe. When there was nothing left to consume, families began to abandon farms. They solicited charity from relatives, wandered the countryside as beggars, or poured into the cities. Entire villages were deserted. Farmers became laborers. The year 1316 was the worst for cereal crops in the Middle Ages. A Flemish observer wrote: "The people were in such great need that it cannot be expressed. For the cries that were heard from the poor would move a stone, as they lay in the street with woe and great complain, swollen with hunger."

Churches and almshouses saved countless lives by cooking up pottages with whatever nourishment they could find and ladling it out to the starving. Abbeys and monastic houses reported that the victims came from almost all social lines, though the majority were poor. In Tournai, "men as well as women from among the powerful, the middling, and the lowly, old and young, rich and poor, died daily in such great numbers that the air was almost wholly corrupted" by their stench. In Holland, "rich and poor alike found it impossible to secure food; they roved along the roads and footways and laid their starved forms down to die," Lucas wrote.

Disease spread quickly through weakened bodies living in damp conditions. In the rains, dead bodies "began to decompose at once." In cities from Erfurt in Germany to Colmar in France, authorities dug huge trenches to bury thousands of dead townsfolk. Scholars estimate the Great Famine of 1315–1322 killed some 3 million people, roughly 10 percent of those living in rain-swamped northern Europe in the early fourteenth century. It must have seemed like the end of the world. But the next Horseman of the Apocalypse to come galloping in on bad weather, the Black Death, would kill numbers much larger.

In fourteenth-century letters and chronicles, tales of the disease then called "the great mortality" or "pestilence," later christened the Black Death, always seemed to begin with strange rain. A musician in the papal court of Avignon wrote home to Flanders in 1348 that when the great mortality began: "On the first day it rained frogs, serpents, lizards, scorpions and many venomous beasts of that sort. On the second day thunder was heard, and lightning flashes mixed with hailstones of marvelous size fell upon the land, which killed almost all men, from the greatest to the least. On the third day there fell fire together with stinking smoke from the heavens, which consumed all the rest of men and beasts, and burned up all the cities and castles of those parts." (The musician's patron would soon die of the pestilence.)

In fact, real rain foreshadowed the Black Death much as it had the

Great Famine. When normal weather patterns returned in the 1320s, northern Europe arose from famine and saw a relatively prosperous few decades. French historians refer to the period as the *monde plein*. But the "full world" did not last long. The early 1340s brought the heaviest summer rains and flood disasters of medieval times, wrecking great bridges, washing away towns, and wiping out crops. Then, in 1344, extreme rains gave way to extreme drought. Localized famines set in; scholars believe these and the Great Famine helped make the Black Death as severe as it was, weakening the population. The stress of childhood hunger would have created a lifelong susceptibility to disease.

What we know as bubonic plague came riding into Europe not on slithering serpents, but on a tiny insect, the flea, which came riding on a rat, which came riding on Mongol supply trains traveling from southern China across Eurasia. The bacterium, *Yersinia pestis,* lives on wild rodents and spreads among them via fleas, which sometimes jump to humans. If *Y. pestis* reaches the lungs, it turns into the deadly pneumonic plague. When a victim coughs, infected droplets spew into the air, infecting the next person and the next and the next. Once infected, a person can die within a day.

The Black Death entered—and might have even been catapulted—at the besieged port of Caffa (now Feodosija, Ukraine). Thousands of Tartars were entrenched there, having surrounded the city walls, behind which Christian merchants were taking cover. An Italian notary named Gabriele de' Mussi wrote that a horrid and fast-spreading disease struck the Tartars, who began to die en masse. Medical attention was useless; they would perish soon after "coagulating humours" showed up in their armpits and groins. Stunned and stupefied by their misfortune, de' Mussi wrote that the final survivors made a last-ditch effort to smite their enemies: "They ordered corpses to be placed in catapults and lobbed into the city in the hope that the intolerable stench would kill everyone inside." Soon, almost everyone who had been near "fell victim to sudden death after contracting this pestilential disease,

as if struck by a lethal arrow which raised a tumor on their bodies."
Some modern scholars discount the claim. Others who have examined
it, such as the University of California biological weapons expert Mark
Wheelis, believe it was a plausible and spectacular attack of biological
warfare.

From Caffa, fleeing Genoese ships unwittingly carried the rats,
with their fleas and their *Y. pestis,* to Genoa and Pisa. At least 35 percent of Genoa's population died in the first sweep of Black Death,
which then spread across western Europe. The population of Paris was
cut down by at least two-thirds between 1328 and 1470. The plague
entered Britain through several ports, including Bristol, "where almost
the whole strength of the town perished." By 1349, it had crept north
to Scotland, where "nearly a third of mankind were thereby made to
pay the debt of nature." In the end, an estimated third of the European
population succumbed to bubonic plague, with regional death rates
between 10 and 60 percent. No event in European history, including
war, has left such devastation.

And what of the rain? Only recently have researchers begun to
unravel the extent to which the prevalence of *Y. pestis* increases with
warmer temperatures and wetter weather. Biologists using tree-ring
data to reconstruct historical climate find that the years of Black Death
in Europe in the fourteenth century were both warmer and increasingly wet. They found the same during a third pandemic, which began
in Asia in the mid-nineteenth century and killed millions of people
in China and India. They find the same today. A global consortium
of scientists studying *Y. pestis* in great gerbils in Central Asia, where
human plague is still reported regularly, found its prevalence increases
with warmer springs and wetter summers.

The story is never so simple, as severe drought can spur plague,
too—that's because it sends rodents on the move. During the American yellow fever plagues of the nineteenth century, heavy summer rains
could spur outbreaks, as standing water created breeding grounds for
the yellow fever mosquito, *Aedes aegypti.* At the same time, a good,

hard rain was the only hope of washing away the sewage, rotting dead animals, and other filth that helped spread the disease in cities such as Memphis in the days before sanitation and plumbing.

Rain can be a Janus; at once the face of salvation and despair. In the days before meteorological science could help explain it, extreme rain also brought damnation—as the storm-weary began pointing fingers at one another, looking to lay blame.

I n late August 1589, a dozen of the fittest ships in the Danish fleet set across a tempestuous North Sea to carry a fourteen-year-old princess bride to her new husband and new home. King James VI of Scotland had seen the fair Anna of Denmark only in a miniature portrait. He'd arranged a marriage by proxy in her country to avoid a union with Catherine of Navarre. Following her wedding-sans-groom in a palace by the sea, Anna boarded the ship of the Danish admiral Peter Munch, charged with delivering his royal passenger to her Scottish kingdom.

In the crossing, the ships met ordinary storms. When they had nearly reached their destination, an extraordinary gale flew at them from the coast. Twice, they came within sight of Scotland's cliffs, and twice, they were pushed back by a phalanx of rain and "baffling winds" that ultimately blew them all the way to Norway. Munch thought the tempest uncommonly fierce, even for the North Sea. So much so, he began to believe "there must be more in the matter than the common perversity of winds and weather."

He believed the rains and winds were being conjured up by witches.

Munch attempted a third approach. Yet another squall roiled up, this one worse than the last. "The whole fleet was dreadfully tossed," none more severely than the admiral's ship carrying the bride-queen. The dastardly winds snapped a cannon from its moorings and hurled it across the deck, where it killed eight Danish soldiers before young Anna's eyes and "very nearly destroyed her."

The gale kicked Munch's sinking ship north into a Norwegian sound. There, an unseasonable ice storm set in, forcing Anna, Munch, and his surviving crew to hunker down in a miserable village that "produced nothing eatable." Anna wrote distressed letters describing the storms and her near-drowning to her new husband, delivered by a young Danish sailor who agreed to cross the wintry seas for the cause. In what historians describe as the single brave act of his life, King James stocked royal ships with delicacies, meats, wines, and hundreds of attendants, and joined the rescue party to fetch his bride from her icy outpost. The troupe embarked October 20 and met a solid month of more horrid weather, not reaching Anna until November 19.

After an awkward first embrace, the couple came together with "past familiaritie and kisses," wrote the Scottish diplomat Sir James Melville, relieved to be united against the diabolical weather still stirring between the Norwegian mountains and the North Sea. Conditions remained too uncertain to sail to Edinburgh for a planned royal wedding. So they held a makeshift ceremony in Oslo, then a third wedding in Denmark, after being led south by four hundred troops sent by the king of Sweden to guide them through his snowy dominion.

The royal couple's return voyage to Scotland confronted yet more "unnatural weather," with the English navy guiding them into the Firth of Forth through a treacherous mist. They finally arrived in Edinburgh in May 1590, when Anna became Queen Anne in an elaborate coronation.

King James had been skeptical of the witchcraft frenzy sweeping Europe. Now, having lived through the freakish gales, he could no longer deny the evidence. The king became as convinced as Admiral Munch and many others in the storm-battered populace that witches had brewed the worst weather in memory to keep the new queen from ascending her throne.

An aging midwife, Agnes Sampson, and a local schoolmaster, James Fian, would pay unspeakably for the squalls. They were among thousands of accused witches tortured, garroted, hanged, or burned for the devilish rains, snows, freezes, floods, harvest failures, sickness, infertility, livestock epidemics, and other miseries that plagued Europe in the years between 1560 and 1660. The worst of the witchcraft persecutions line up with the worst decades of the Little Ice Age.

When Americans think of witches and witch trials, our minds turn to Salem in Massachusetts, home to the severest persecutions in the English colonies, with 185 people accused and 19 eventually executed for witchcraft. Salem's notoriety is due to the city's obsession with marketing and branding itself as the world's witch capital. Historians who study witch hunts, trials, persecutions, and executions define Salem as a "small panic" compared with the witch hysteria that swept Europe in the stormiest years of the Little Ice Age.

It is hardly a history to crow about on the Web and in the wax museum. The German historian Wolfgang Behringer has calculated at least 50,000 legal executions for witchcraft in Europe, half of those within the boundaries of present-day Germany. About 80 percent of the victims were female, explanations for which stretch to the story of Eve and its assumption that women were more susceptible to the Devil's temptations.

Like the spread of humans during dry times in Africa, Behringer finds climate just one of many answers for the rage of witch-hunting in Central Europe. But the crime of witchcraft began its rise in the fourteenth century, running in parallel to the rise of the Little Ice Age. A German woodcut from 1486 shows a sorceress conjuring enormous chunks of hail. A frontispiece from a 1489 pamphlet called *Weather Magic* depicts two hags adding ingredients including snakes and chickens to a tall cauldron, as a storm bursts overhead. A colored Swiss painting from 1568 portrays a dance around the Devil, with witches in the background brewing up a storm that runs straight from cauldron to sky.

Persecutions reached their peak during the worst years of the climate extremes, in the decades before and after 1600. The crime disappeared from the penal catalogs after the end of the age—when the sun emerged along with more enlightened explanations for the weather.

The western coastal fringes of Europe, which enjoyed more temperate weather and less vulnerability to famine, saw far fewer witchcraft accusations. Suspicions were rampant in densely populated Central Europe, which was also hit by the greatest climate extremes. It happened that people who lived with the weakest infrastructure and farmed some of the worst soils also faced the most extreme rains and the harshest winters of the Little Ice Age.

Superstition had it that, in addition to conjuring storms, witches could sow disease, kill children, and make men, women, and farm animals impotent. The accusations reflected the miseries people faced in the Little Ice Age: childlessness, disease, sudden death of children, livestock epidemics, harvest failures, late frosts, freak hailstorms, and extreme rains. "The idea that misfortunes of this kind happened by chance was alien to many Europeans of that period," says Behringer. "Witches were the scapegoats that people needed to explain the disasters of the age."

Beginning around 1580, reports of witch-hunting and executions no longer referred to individuals, but to mass round-ups. Published that year in southwestern Germany, a pamphlet titled *Two Newspapers, What Kind of Witches Were Burned* gives a hint of the accusations and scope, detailing 114 executions of witches who "had mainly confessed to having caused damage to cash crops like grain and grapevines by hail and thunderstorms, as well as the laming and killing of children," writes Behringer. In 1582, a similar sheet reported "the devastating storms in August that had destroyed grain and grapevines, and the subsequent burnings: In the Landgraviate of Hessen, where ten women were burned; a small village in the Breisgau, where thirty-eight women were burned; the small town of Turkheim in the Alsace, where forty-two women and a male ringleader were burned; and Montbeliard, where forty-four women and three men were convicted of weather-making and executed on 24 October."

In much of Central Europe, historians have found that the persecutions grew from the grass roots: Villagers harangued reluctant local courts and prince-bishops to do something about the foul weather by rounding up storm makers. But the witch hysteria and executions in Scotland and England debunk the notion that belief in witchcraft was characteristic of primitive, uneducated people such as peasants and serfs. In Scotland, the zeal for witch-hunting came straight from the top—from the same King James who gave us the King James Bible.

So convinced was the king that the tempests of 1589 and 1590 were a murder plot to keep him from his queen, James became personally involved in the investigation, interrogations, and trials of the mortals believed to have worked with Satan to conjure the storms. Known as the North Berwick Witches, they were rumored to have been meeting late in the night at the kirk, or church, at North Berwick, about twenty miles from Edinburgh on the Firth of Forth.

Agnes Sampson's name quickly surfaced as ringleader. A renowned midwife and healer known as the Wise Wife of Nether-Keith, her work smacked of witchcraft. She made potions for healing and midwifery. She sold love charms and other talismans. After humiliations and tortures that included having her entire body shaved, being made to wear a witch's bridle, sleep deprivation, and being probed in the genitals with a pin, Sampson confessed to conjuring the storms that had hindered the union of James and Anne.

Eager to appear the deep-thinking skeptic, King James jumped up at that dramatic point in Sampson's interrogation and called her a liar. Even more eager to stop her torture, Sampson asked to speak with James privately. She convinced him of her witch's power by telling him the secrets he had whispered to his new wife on their wedding night in Oslo. Then she told James that the Devil hated him and wanted to see him drowned by storm. She explained that Satan considered the king his chief opponent in all the world.

No words could have rung so true to the self-important and deeply religious king. James now saw himself as the avenging knight of the Christian faith. Sampson was only one of many who would pay the price. She was burned at the stake in winter 1591 at Edinburgh Castle.

Trudging up the half-crag, half-castle in a bitter winter wind, it is not hard to imagine Agnes Sampson's January execution. The fortress is built into a jagged volcanic formation, 440 feet above the sea. The 1590s are believed to have been the coldest decade of the sixteenth century. Crowds would have climbed the hill in freezing winds to watch Sampson strangled in a garrote. The townspeople would have lingered late into the night to watch the fire for the hours it took to burn her body to ashes.

The story, at least as the king and the courts saw it, was printed in a pamphlet widely distributed at the time, *Newes from Scotland*. Its woodcuts show a storm ravaging the king's ship on the Firth of Forth, with women huddled around a boiling cauldron on shore. Sampson had provided her interrogators with a storm recipe that included joints or knucklebones from corpses, and a cat cast into the sea. She said the witches carried out the plot by sailing up the firth in magic sieves, then calling up the storms.

Newes describes the torture of a maidservant, who endured graphic agonies before she finally blamed Sampson and others, including the schoolmaster James Fian, for the storms. Dr. Fian never confessed, even though his legs were totally crushed in *bootes*. King James and his council decided to burn him at the stake anyway, as an example, *to remayne a terror to all others hereafter.*

King James remained so concerned about the threat of witchcraft to himself and Scotland that he wrote a treatise on it, *Daemonologie.* Published in 1597, the book was meant to "resolve the doubting of many," so that people would believe in the "fearefull abounding at this time in this Countrey of these detestable slaves of the Devil, the Witches or enchaunters."

When Queen Elizabeth I died in 1603, he was crowned King

James I of England and ruled both countries until his death in 1625. At the time he ascended the British throne, a witch could be hanged in England only if it were proven the witchcraft caused a death. One of James's first acts was to strengthen England's Witchcraft Act, requiring hanging for *any* witchcraft confessed or proven. Witch hunts and trials continued unabated during his reign. They started to trail off only in the early eighteenth century.

The drama that began with the North Sea storms of 1589 and ended with the witch burnings of winter 1591 may feel vaguely familiar. It brings to mind the "shipwrecking storms and direful thunders" of one of the most powerful plays in the English language. William Shakespeare's *Macbeth* opens to "Thunder and lightning," according to the bard's directions. "Enter three Witches." The opening lines:

> **FIRST WITCH:** *When shall we three meet again*
> *In thunder, lightning, or in rain?*

Shakespeare wrote *Macbeth* shortly after James became king of England. The playwright and his troupe, the Lord Chamberlain's Men, had enjoyed independence and quiet fame during the reign of Queen Elizabeth. That would change within weeks of James's ascension. The new king made Shakespeare's company his own and called it the King's Men. For Shakespeare, it would mean unprecedented exposure and success, but also new worry about pleasing the crown. He began writing with James in mind, "burrowing deep into the dark fantasies that swirled about in the king's brain," writes the American literary critic Stephen Greenblatt. England had carefully watched James's obsession with witchcraft. Satan's stormy plot against James and Anna surely informed *Macbeth*'s seafaring witches traveling to Aleppo in a sieve, and some of the grim ingredients tossed in their cauldron.

Aside from appealing to his paranoid king, Shakespeare had come to understand the sweeping power of rain in the human drama. His comedies were often sunny. But he relied on storms as foreboding signs, symbols of chaos, and revelations of character in not only *Macbeth*, but *King Lear*, *Othello*, *Romeo and Juliet*, *Coriolanus*, and, of course, *The Tempest*.

From Shakespeare to the Rig Veda, the rains of history influenced our stories of origin and those of end—a mythical river coursing through the Indus Valley's Harappan civilization, storm gods buried under the dry sands of abandoned city-states.

In the drought of 4,000 years ago, the Harappan cultures did not die out entirely. Rather, the lamps of the largest cities in the ancient world dimmed as people moved out of the Indus Valley to the south and east in search of their lost rain. Archaeologists have found that the later Harappan settlements are rural, and more numerous in rainier regions at the foothills of the Himalayas and along the Ganges River in northern India.

There, the religious traditions of Hinduism would evolve to turn rivers into goddesses and make a beloved god named Krishna blue—for the color of storm clouds.

Rain would influence the evolving religions in deep but starkly different ways. It often depended on whether the faithful danced to the rhythms of the quenching monsoons—or marched to the beat of punishing aridity.

THREE

PRAYING FOR RAIN

Mid-nineteenth century, when Texas was still a rowdy republic, a popular judge and lay preacher named Robert McAlpin Williamson was known by the nickname "Three-Legged Willie." As a teenager, he'd been stricken by tubercular arthritis that confined him to bed for months and permanently paralyzed his right leg, which bent back at the knee. His custom pants had three holes: one for his left leg, one for the back-pointed one, and another for the wooden peg he wore below his right knee.

Williamson read so voraciously during his illness that he emerged a prodigy and a crack lawyer—admitted to the bar around age nineteen. His disability stopped him from nothing, including becoming an expert horseman and marksman, a Supreme Court justice and congressman for the Republic, and one of the first majors in the Texas Rangers.

As a brilliant orator and trusted leader, Williamson also had another honor to uphold. He was the go-to preacher when it came time to pray for rain.

Texas in the 1840s was best known for storied battles among the Rangers, the Comanche, and the government of Mexico for the soul of the Lone Star State. But Texans were also fretting over the soul of the skies. Tree-ring researchers who read the history of rain in ancient bald cypress know that 1840 to 1849 marked one of the worst drought decades in Texas history. Settlers described conditions straight out of the Bible: invading grasshoppers, dust storms, wildfires, dead cattle. Creeks, springs, and rivers dried to mud. Bleached bones littered the land. Wheat failed. Corn grew to nubs.

A typical Williamson prayer:

O Lord, Thou Divine Father, the supreme ruler of the Universe, who holdest the thunder and lightning in thy hands, and from the clouds givest rain to make crops for thy children, look down with pity upon thy children who now face ruin for the lack of rain upon their crops; and O Lord, send us a bounteous rain that cause the crops to fruit in all their glory and the earth to turn again to that beauteous green that comes with abundant showers. Lord, send us a bounteous one that will make corn ears shake hands across the row and not one of these little rizzly-drizzly rains that will make nubbins that all hell can't shuck.

For those who settled in Texas, the risk of drought would hang like red dust clouds over the plains. A century and a half after Three-Legged Willie's time, in 2011, Texas burned with more than eight thousand wildfires brought on by a drought to rival the 1930s Dust Bowl. At Texas A&M University, atmospheric scientists said global warming was making the hellish conditions worse. The governor of Texas, Rick Perry, was skeptical of the professors. But he could sink his black-leather boots into prayer. Perry called upon his fellow Texans to join him in three days' prayer for rain. Whereas, he wrote, Texas had received no rainfall for nearly three months; whereas fire had engulfed more than 1.8 million acres of the state and destroyed four hundred homes; whereas crops and businesses had failed . . .

I, Rick Perry, Governor of Texas, under the authority vested in me by the Constitution and Statutes of the State of Texas, do hereby proclaim the three-day period from Friday, April 22, 2011, to Sunday, April 24, 2011, as Days of Prayer for Rain in the State of Texas.

At the time, Perry had his eye on the U.S. presidency. He showed off his boots, hand-stitched with the slogan of the Texas Revolution, "Come and take it." He gave speeches in fluent Spanish to blue-leaning Mexican Americans. Ultimately, his brand of traditionalism did not win wide enough appeal. His rain refrain brought him criticism as a biblical bully, and for "trying to co-opt the most important three days of the Christian calendar"—Good Friday through Easter Sunday.

But Perry's prayer for rain bowed to tradition much older and broader than Texas, or even Christianity. In the arid American Southwest, the Indian Rain Dance remains an exaggerated cliché of the reality that rain prayer has been a part of daily life since ancient times, clear from the cloud designs on Ancient Puebloan water pitchers or the frog bracelets and pendants worn by the prehistoric Hohokam in what is now Arizona. Rain has been woven so deeply into the spiritual life of many natives that it became a name, a clan affiliation, or a personal symbol—like the rain cloud signature of potters and jewelers in the Hopi Water Clan.

Whether Three-Legged Willie in the 1800s, Governor George W. Bush of Texas in 1999, or Governor Sonny Perdue of Georgia in 2007, Christians have long prayed for storms to relieve parched land in dry times. Jews and Muslims do as well. Jews around the world pray for rain each year on the eighth day of Sukkot, the pilgrimage festival that celebrates the harvest. The cantor dons a white robe and recites the special rain prayer, *tefillat hageshem,* to mark the beginning of rainy season in Israel. In Islam, Prophet Muhammad himself performed the rain prayers when he was alive. Raising his hands to the sky, with his back to the crowd, he would turn his cloak inside out. Today, Muslims communally perform the prayer, reciting the *salatul istisqa,* turning their outerwear inside out, and raising hands.

In an unprecedented gesture of unity during a record-dry recent autumn in Israel, Christians, Muslims, and Jews gathered together in a valley between Jerusalem and Bethlehem to pray for an end to the drought. Rain can be a powerful unifying force. As it brings modern

Abrahamic faiths together in joint prayer, rain also connects them to the birth of their religions in the searing ancient deserts of the Middle East.

More than four thousand years before Three-Legged Willie prayed for hand-shaking corn and Governor Perry asked God to douse wildfires, one of the earliest gods known to have been worshipped by humans was a deity of storms and rain. In Mesopotamia, the lightning-bolt-wielding rain god stood balanced on the back of a galloping bull, riding through a wild tempest in the sky. He was known as Iškur by the Sumerians, or Adad by the Akkadians, who referred to storm clouds as Adad's "bull calves." (The rain god Iškur/Adad was in some traditions son of Enlil, the drought-and-flood-making misanthrope blamed for desiccating Mesopotamia.)

In pollen grains and deep-sea cores, we saw how geologists found evidence for a climate transition from moist to drought-prone around the time humans put down spears and picked up hoes. In that same period, archaeologists note a cerebral shift among worshippers in the earliest city-states. Where the hunter-gatherers had prayed feverishly for fertility, the agriculturalists upped the spiritual ante on rain. Mother-goddess artifacts are abundant in the earliest Mesopotamian cultures and others during Neolithic times—busty clues to a culture focused on procreation. Rain and storm gods were around then, too, but it wasn't until the shift to agriculture and urban living that scholars find increasingly urgent references in images and texts to male storm gods such as the bull-riding Iškur/Adad.

In regions such as Upper Mesopotamia where agriculture depended more on rain than on irrigation, storm gods ranked as the most prominent of all gods. Some were considered divine kings who ruled over the other gods and could even bestow kingship on humans. Still thought of as paeans to fertility, they came to conjure life along with rains when they felt appeased—drought and floods with infertility when angry.

Bulls were a common rain-god motif not only because of their thundering run, but as symbols of masculinity and sexual power. Rain goddesses were around, too, but often as naked escorts of the storm gods.

The male rain god would endure in the driest cultures. As late as the sixteenth century, Spanish chroniclers described the Aztecs sacrificing children to their rain god, Tlaloc. Scant archaeological evidence substantiated the stories until recently, when researchers found the skeletons of thirty-seven children and six adults carefully laid out in what appears to have been a single ceremony at a temple in Tlatelolco, in what is now Mexico City. The remains, which include tiny infant bones folded into urns, date to a time of drought-induced famine in the mid-1400s.

The Aztecs believed many smaller gods lived in the hills and mountains and acted as helpers to Tlaloc, and that these little assistants had hands-on responsibility for rain. The temple at Tlatelolco was devoted to one of the wee rainmakers, Ehecatl-Quetzalcoatl. When molecular anthropologists analyzed the victims' DNA, they found that most of them were younger than three. All whose gender could be determined were male—Ehecatl-Quetzalcoatl personified. The Aztecs had tried to please the elfin rain god with the closest likenesses they could find.

Many religions and cultures came to view the rain, itself, as male—though Native Americans consider driving rain male and soft rain female; both equally vital to sate life and landscape. When the Hebrew God was creating Earth, he divided the waters in the heavens above from those on land below. Jewish tradition identifies the upper waters—the rainfall—as male, and the lower—lakes, rivers, and springs—as female, citing the line in Isaiah, "Let the Earth open to receive, that it may bear the fruit of salvation . . ." In Sanskrit, the word for rain, *varsha*, is derived from the older *vrish*, which means not only "to rain," but also "to have manly power" and "generative vigor." Hindus consider rivers female, and sometimes describe those swollen with monsoon rains as pregnant.

Some cultures made a more literal link between rain and semen;

farming couples took to making love in the fields to induce rain. Others sent nude women into the crops to sing ribald songs to the rain. Australian Aborigines bled tribesmen they considered rainmakers, sprinkled their blood over other men in the tribe, then made all the participants avoid contact with their wives until the rains arrived. The Australians also attributed rain-giving powers to foreskins removed during circumcisions. They would put the skins away for safekeeping, out of sight of any women, to break out in case of drought.

The Sumerian Iškur and his Akkadian counterpart Adad were ambivalent rain gods, foreshadowing the alternately benevolent and punishing God of monotheism. They could intervene to help humankind by bestowing rain for crops, or unleash their wrath in a drought or monster flood, usually for some moral reason. (Other storm gods became known for arbitrary destruction, such as the Yoruba Shango, the Polynesian Tawhiri, and the Japanese Susanoo, so unruly he's banished from heaven.)

One cuneiform tablet describes Iškur as "clothed in a frightful radiance, who by means of his thunder gathers the thick clouds, who opens the teat of heaven, who makes produce and abundance plentiful everywhere." The ancient stories of Iškur and Adad also offer a desolate glimpse of long-term drought; the stripped soil is evident in the Mesopotamian epic *Atrahasis*: "Above, Adad made scarce his rain. Below, the fountain of the deep was stopped, that the flood rose not at the source. The field withheld its fertility. A change came over the bosom of Nisaba; the fields by night became white."

I n the Old Testament, Abraham makes a pact with God to give up all this sort of idolatry and be loyal only to Him. Abraham promises to leave northern Mesopotamia and its polytheism. In return for Abraham's faith, God will take care of him and his descendants in the pastoral, rain-fed land of Canaan.

Several hundred years later, starving in an unrelenting drought, many of these descendants lose their faith and flee for prosperous Egypt. At first, they thrive. But in time, they're enslaved and forced to build cities along the Nile Delta. Eventually Moses releases them from bondage and leads them back toward Canaan and freedom.

In the exodus that follows, here's the climax and its climatic twist: God wanted his people back where he could control them by means of rain and drought. As Methodism's founder, John Wesley, explained nearly three hundred years ago in his still-popular *Notes on the Bible*, God was careful to settle his people in Canaan rather than Egypt, "not in a land where there were such rivers as the Nile, to water it and make it fruitful, but in a land which depended wholly upon the rain of heaven, the key whereof God kept in his own hand, so that he might the more effectually oblige them to obedience."

The newly settled Israelites anguished endlessly about the availability of rain for their farms and pastures, and the timing; early showers were crucial to germinate seeds and sprout new crops. God promises not only rain, but rain at the right time: "And it shall come to pass, if ye shall hearken diligently unto my commandments which I command you this day, to love the Lord your God and to serve Him with all your heart and with all your soul, that I will give you the rain of your land in his due season, the first rain and the latter rain . . ."

Rain is God's way of keeping us honest, according to the tradition, for the promise is immediately followed by the warning: "Take heed to yourselves, that your heart be not deceived, and ye turn aside and serve other gods and worship them; and then the Lord's wrath be kindled against you, and He shut up the heaven, that there be no rain and that the land yield not her fruit, and lest ye perish quickly from off the good land which the Lord giveth you."

As God told Job, only God is the father of rain and only He can create it. (Tell that to the geoengineers!) Only He can bestow it as a blessing or take it away as punishment. Idolatry, bloodshed, or lawlessness could all bring the wrath of drought. With the notable exception of the Great Flood, God's rain wrath takes the form of drought far

more often than deluge. The Hebrew Bible, or Christianity's Old Testament, is full of tales of God starving crops of rainfall; sending it to
one city but not another; withholding rain for months or years.

But it is likewise full of stories of rain as God's blessing—His "good
treasure" falling upon Israel and all of Earth: "The Lord shall open
unto thee His good treasure, the heaven to give the rain unto thy land
in his season, and to bless all the work of thine hand."

Indeed, in a host of religious traditions, from Allah fracturing
clouds into raindrops in Islam to Buddhist rain-cloud kings, rain is
among the most important blessings possible. In Judaism, rainfall is
said to be one of life's greatest events, greater even than the giving of
Torah. As Rabbi Tanchum bar Chiyya put it in the third century, "the
giving of Torah was a joy to Israel, but the falling of rain is a joy for all
the world."

Religions hold a mirror to the history of humans and their complicated worlds, including their beliefs and perceptions about climate.
The monotheism of Christianity, Islam, and Judaism all grew out of
the arid sands of the Middle East. Some historians trace monotheism to agriculturalists in these dry lands looking to the skies for life-
giving rainfall. Most of the polytheistic religions were born in the
soaking monsoons. Some scholars speculate that as people evolved
their belief systems in radically different climate conditions, they took
radically different approaches to interacting with God, nature, and
one another. "In the wilderness of the desert, where life struggles to
survive, it would seem logical that a divine being would be responsible
for the creation of living things out of nothing, and that in due course
time and life will end in a final day of judgment," writes the geoscientist Peter Clift, who studies the Asian monsoon and its human
impacts. "In contrast, in the forested land that has grown under the
influence of summer monsoon rains, life is everywhere and abundant.
Tropical forests teem with life and the cycle of birth, life, and death
are endlessly replayed, resulting in a theology that does not emphasize
a beginning or end of creation."

———

The faithful pray for rain in every climate zone, for even the rainiest parts of the world can sometimes face drought. But it is often the wettest, monsoon-drenched regions in which the rains and rivers themselves have achieved immortality. This is especially true in Asia. In India, the nation's estimated one billion Hindus consider the rain-fed Ganges River holy. And the lives of gods including revered Krishna are intimately tied to rain. Krishna's skin is storm-blue, and his name means "dark as a storm cloud." Rain follows him from the day of his birth to a royal family in Mathura during a terrific storm. The tempest helps obscure a ruse when his father secrets Krishna across the Yamuna River (the largest tributary of the Ganges) to switch him with the newborn child of a cowherd couple so he won't be murdered by Mathura's wicked ruler.

In one of Hinduism's best-known stories, as a young man, Krishna convinces the people of the region to stop worshipping Indra, a rain god who is king of the gods as well as the storms. Krishna suggests his cowherd friends worship Mount Govardhana instead; Krishna will become the mountain and receive their offerings. Indra flies into a rage and sends down angry rains. Then Krishna lifts Mount Govardhana and holds it over the cowherds as a gigantic umbrella. The image of blue Lord Krishna balancing the mountain umbrella effortlessly with his finger, sheltering his happy companions, is the subject of some of Hinduism's most significant artwork. For centuries it has been carved into temples and stone walls in bas-relief, embroidered into textiles, and painted with whimsy and bright colors, a menagerie of animals often tucked into the mountain crevices, rain falling harmlessly around.

Over time, Indra becomes far less powerful, and less popular, than Krishna. And worship of rain and rivers centers around the Ganges. The 1,570-mile Ganges—the world's most heavily populated basin and sadly one of its most polluted rivers—rises in the Himalayas and flows south and east across the northern Indian plain into Bangladesh. There, the river returns the monsoon rains to their birthplace, the Bay of Bengal. Physically, the Ganges is a pilgrimage for Hindus. Personified, it is the goddess Ganga. A beautiful woman who sports a fish tail instead of legs, Ganga perches on Makara, a crocodilian water monster. In her right hand is a water lily, a symbol of rain and fertility. In her left, she holds a water pot. Hindus celebrate the Ganges in all sorts of rituals and festivals, including Kumbh Mela, which draws the largest single gathering of humanity in the world to plunge into a ritual bath, their faces joyous, colorful saris soaked to the skin. In 2001, the Indian government estimated that 70 million people congregated by the banks of the Ganges to bathe in its sacred waters.

Come summertime, when the great monsoon rains blow in from the Indian Ocean, hundreds of monsoon festivals likewise draw pilgrims or partiers to cities and villages throughout India. On the southern banks of the Brahmaputra River in the far eastern state of Assam—one of the rainiest places in the world and among the most sacred destinations in India—travelers descend upon the Kamakhya Temple for the Ambubachi Festival. As the monsoons swell the Brahmaputra, they are said to bring on the annual menstruation of the temple's presiding goddess. The hilltop temple shuts down for her three-day period. A crush of devotees waits outside, and when the temple doors open, they make a mad rush to receive a small bit of cloth moist with the menstrual fluid—infused with all the power of the monsoons, fertility, and other blessings.

As the goddess Ganga and the Ganges River are central to the faith of Hindus, so remains the lost Saraswati River that is said to have flowed through the ancient Indus Valley Harappan civilization. Mythology surrounding the Saraswati was committed to memory and passed on in stories and songs for generation upon generation before being written down in the Vedas. Along the way, Saraswati, too, became a goddess—patron of arts and education.

eonard Woolley, the archaeologist digging at Mesopotamia's Ur for a Genesis-scale flood, once said that "we ought to assume that beneath much that is artificial or incredible, there lurks something of fact." Woolley was not trying to make history out of legends, he explained. Rather, he was trying to find history within legends. Surely there were truths waiting to be discovered in the greatest rain story of all time.

efore the nineteenth century, westerners knew the story of the Great Flood only from the book of Genesis in the Hebrew Bible. The tale is as familiar as it is vengeful: Ten generations since Adam, humankind has become terribly wicked, "for all flesh had corrupted their ways on Earth." God is sorry to have ever created the place, and decides to wipe out everything and start fresh. But he tips off one imperfect yet pious soul, six-hundred-year-old Noah. "I will cause it to rain upon the earth forty days and forty nights," God tells him, "and every living substance that I have made will I destroy from off the face of the earth." God goes on to explain how Noah should build a wooden ark and save his family, along with a pair of each living creature. He lets loose the flood in rains from the heavens above and a surge of waters from belowground, until it drowns even the highest mountains. The disaster lingers interminably; in one account for 150 days, in another, a full year. Finally, Noah and his zoo crew float in an eerie, silent devastation until the ark comes to rest in the mountains of Ararat. Noah sends out a raven that never returns, then a dove that comes back with an olive leaf, a sign the waters are receding. Noah builds an altar and makes some burnt offerings to God. At the end of the saga, God promises that, despite our irretrievable evil, he'll never again set out to destroy Earth and its living creatures, at least not with a flood. Rainbows exist as a sign of this pact: a reminder of "the everlasting covenant between God and every living creature of all flesh that is upon the earth."

In mid-1800s London, a young engraver named George Smith who was enthralled with Noah and other tales of the Old Testament (and also rocked a fabulous Noah beard) became obsessed with the antiquities turning up in the ancient Near East. British archaeologists were digging up stunning artifacts from the deserts of Iraq. Nineveh, flourishing capital of the Assyrian Empire, materialized straight off the pages of Genesis. Archaeologists had found the library of Ashurbanipal, king of Assyria in the sixth century B.C. Smith worked for a firm of bank-note engravers near the British Museum, where cryptographers were trying to piece together and translate thousands of cuneiform tablets and fragments from what was the oldest surviving library in the world. He began spending all his extra time hanging around the museum, shelling out whatever money he had on books to teach himself cuneiform and Assyrian. His instincts for the material impressed the museum's scholars. They soon gave him a job, essentially piecing together two-thousand-year-old jigsaw puzzles. Smith had to examine and sort through thousands of marked bits and try to match them with their originals. He loved the work so much that he would become furious when the museum had to close due to London's fogs; it did not yet have artificial lighting.

Smith was "a highly nervous, sensitive man," according to his museum coworkers. Over a period of weeks in 1872, he was beside himself waiting for one of them to return to work to clean a limestone-like deposit that had solidified on an intriguing corner of tablet.

Only four inches wide, the hunk of clay would change Smith's life. When the offending material was finally scrubbed off and he could make out the tiny scratches, what he read felt very, very familiar. He wrote of the moment, "my eye caught the statement that the ship rested on the mountains of Nizir, followed by the account of the sending forth of the dove, and its finding no resting-place and returning. I saw at once that I had here discovered a portion at least of the Chaldean account of the Deluge."

Smith held in his hands an account of the Great Flood written at least a thousand years before the Bible's first books. (Not to mention centuries before Homer and his heralded launch of the literary canon.) According to a colleague's written account, Smith set the clay piece on the table, then "jumped up and rushed about the room in a great state of excitement, and, to the astonishment of those present, began to undress himself."

Smith had discovered number eleven of the twelve chapters in the *Epic of Gilgamesh*. The story of the legendary Mesopotamian ruler Gilgamesh and his search for immortality is among the world's earliest surviving works of literature. The piece Smith found describes Gilgamesh's meeting with Utnapishtim, the trusted adherent forewarned of a plan by the gods to send a great flood across Earth. The telling is near-exact to Genesis: the warning and handpicked survivor; his building a wooden ark and saving "all the beasts and animals of the field"; the landing on a mountain; the dispatch of a swallow, raven, and dove to find land; even a final promise that the gods would never again sink humanity into watery chaos.

The flood tale got around the ancient world. The Babylonian version was known to the Canaanites. Fragments of the story have been found in central Turkey, in the royal library of the Hittites, who are believed to have transmitted it to the Greeks. The stories merge in Greek mythology, where the most powerful god, Zeus, was lord of the sky and the rain. In the Greek version, Zeus becomes so disgusted with humankind that he unleashes a tremendous flood to wipe out almost everybody. Only ark-building Deucalion and his wife, Pyrrha, are saved.

The Greeks pass the allegory to the Romans. Their Jupiter, like the Greeks' Zeus and the Hindus' Indra, was king of the gods and also meted out rain; in that role, he was called Jupiter Pluvius. Writing in Rome at the close of the first century B.C., Ovid recounts Jupiter's disgust with the evil deeds of humans—their contempt for the gods, their violence, their lust for slaughter. He decides to wipe them out, which disappoints his fellow gods because . . . who will bring incense to their altars? No worries, Jupiter says, he'll create another race of beings far superior to the first. Jupiter pours on the rain, lets loose the wind, sets the sea god Triton to raise huge waves, and makes the rivers overflow their banks. Most of humanity drowns. The rest starves. By the time Jupiter stops the rain and Triton blows his conch shell to calm the waters, only Deucalion and Pyrrha are alive. They start a new race from stones.

In Hindu lore, Brahma turns into a fish to warn his son Manu—the first man—of worldwide flood. Manu also builds a large boat and gathers seeds from all life. When everything else is wiped out, he makes an offering to the gods, who then produce a beautiful woman, with whom he parents a new race.

For more than a century, the torrent tales from around the globe have sparked endless theories, research expeditions, books, and more than a movie or two. Inspired by Smith's find and other clues, Leonard Woolley at his Ur excavation in the 1920s was convinced that if he could dig deeply enough through time, he could find tangible evidence of a Genesis-scale flood.

Shoveling into the strata below the Sumerians with their art and metallurgy, Woolley ultimately dug down to the first settlers of Mesopotamia, and the time when Ur was still a tiny village. There, he hit waterborne silt. What was clearly a flood deposit went on for ten continuous feet. It had inundated houses and temples. Woolley believed he had discovered an ancient deluge large enough to have wiped out an entire population.

Woolley's announcement grabbed the world in newspaper headlines, radio broadcasts, and movie-house newsreels, not to mention church pews. Not only had Woolley unearthed a lost biblical city, but now he had literal proof of one of the most important stories in the Hebrew Bible. His 1929 book, *Ur of the Chaldees*, became the most widely read book on archaeology ever printed.

Woolley's great flood discovery lived on in the popular imagination for decades. But scientifically it turned out to be a bust. Subsequent excavations at Ur and in neighboring ancient settlements failed to reveal the same silt layer. After decades of increasingly technical probing, archaeologists determined the Ur flood was localized—perhaps only a single breach in a Euphrates River levee.

Could a local flood have inspired such grand epics? Any flood would feel like the end of the world if your neighbors drowned and your community washed away. In Mesopotamia when torrential rains hit alongside spring snowmelt, the Tigris and Euphrates would burst their banks, drowning the region under hundreds of miles of lakes. Archaeologists say an ancient Sumerian city called Shurrupak (Iraq's Tell Fara) was laid waste by flood nearly 5,000 years ago. A Babylonian version of *Gilgamesh* mentions Shurrupak by name. It describes a deluge that wipes out mankind, and a pious king called Ziusudra who overhears from a sympathetic god that the great flood is on its way. Ziusudra builds a huge boat and survives.

To the American geologists William Ryan and Walter Pitman, a river flood doesn't jibe with the drama of the narratives. There would be no warning of disaster, no time to build an ark. For two decades, Ryan and Pitman have built evidence for an actual flood in the Middle East 8,500 years ago that would have brought a cataclysm worthy of *Gilgamesh* and Genesis. They make the case that today's Black Sea was once a smaller and landlocked freshwater lake. Sediment cores and high-resolution imagery of the seafloor reveal once-dry plains. Neolithic people likely settled on the fertile lands to farm and harvest fish. The worldwide climate was still rapidly warming following the last Ice Age, the seas steadily rising. The geologists hypothesize that the oceans rose to a critical point that pushed the Mediterranean Sea through the narrow Bosporus Strait, which divides modern Turkey from Europe—with a daily force perhaps two hundred times greater than Niagara Falls. The waters would have risen ominously day after day and week after week. As villagers realized this was not the beneficent annual flooding that helped seed their crops—but a sea of death—they would have torn apart their homes and sheds to obtain beams and braces for makeshift boats to flee.

Following in Ryan and Pitman's footsteps, a team of scientists and engineers led by the oceanographer Robert Ballard, best known for finding the sunken ocean liner *Titanic* in 1985, began to map the Black Sea floor in search of a lost settlement. Remote underwater vehicles beamed up sonar images revealing both the plains and the shoreline of the ancient lake. Yet so far, no one has been able to find Noah's neighborhood—just some extinct freshwater shells and well-preserved shipwrecks at the bottom of a lonely sea.

Skeptics use the great climate swings of Earth's history—the civilization-crushing droughts and whatever epic deluge may have inspired our flood myths—to argue that the heat-trapping gases of modern life are not to blame for today's global warming and the rise of extreme rains, storms, and floods. In fact, our past climate swings—and the flood myths themselves—give us all the more reason to overcome our differences and confront a new threat with the beams and braces of human ingenuity. The wisdom in the flood stories surely involves heeding the forewarnings from the skies; Noah and our other ark-building heroes have something to tell us about coming together to ride out stormier times. It is what we've done for thousands of years, rain unifying humanity from the deserts to the seas.

II

CHANCE

OF RAIN

FOUR

THE WEATHER WATCHERS

n November 1703, a strange malevolence began to gust through the dark streets of London. Daniel Defoe was walking in his neighborhood on the evening of Wednesday the 24th when "the Wind encreased, and with Squauls of Rain and terrible Gusts blew very furiously." Those winds whipped tiles from the rooftops, snapped limbs and entire trees, and toppled chimneys, one of which very nearly crushed Defoe. Had he been killed, he never would have given the world *Robinson Crusoe,* and one of the most famous literary umbrellas of all time. Nor would he have written *The Storm,* considered by many the first work of modern journalism.

For two more days, the winds swept from the southwest in violent gusts. No one could have imagined they were the outer bands of a storm three hundred miles wide, the largest and most destructive ever to hit the British Isles. When Defoe looked at his barometer on Friday evening the 26th, the mercury was as low as he'd ever seen it. He suspected that "the Tube had been handled and disturb'd by the Children." (Not a bad guess, since he and his wife, Mary, had seven at that time, ranging in age from two to fourteen.)

Defoe was then a poet and pamphleteer who sold his writing by the sheet. He was also just out of prison. For his most recent pamphlet, a satire on the religious intolerance of high-church Anglicans, he was charged with seditious libel. He was fined two hundred marks, locked in an elevated public pillory—the old wooden chokey with holes for head and hands—and jailed for four months. Now bankrupt, he was desperate for paid work to support his family. On the morning of the 27th when the worst of the fury was past and people began to "peep out of Doors," Defoe looked out over the destruction and saw a brand-new genre.

Hardly anyone had slept the night before; "the Distraction and Fury of the Night was visible in the Faces of the People." Many expected their houses to fall in on them, but they could hardly leave, as bricks, tiles, and stones flew through the streets. Unlike any writers of his day, and borrowing from the emerging scientific method, Defoe took detailed notes of his own observations, began to interview eyewitnesses, and set out to gather the grim facts. He visited the Thames to report on the seven hundred or so ships that had been tossed in heaps by the wind. Some of his most vivid reporting came from the Goodwin Sands in the English Channel, where mariners who thought they'd found safety were later washed away.

Defoe's eyewitness account was revolutionary, but he went much further in collecting detailed, personal storm stories from all over England. Journalism was brand new. Nine months before, London's one-page *Daily Courant* launched as the first English-language daily newspaper. Defoe placed ads in the *Courant* and the *London Gazette* and wrote to sources all over England asking for storm stories and particulars. The heart of *The Storm* contains about sixty accounts that Defoe selected, edited, and deemed credible, as "most of our Relators have not only given us their Names, and sign'd the Accounts they have sent, but have also given us Leave to hand their Names down to Posterity with the Record."

Defoe estimated that the storm drowned 8,000 people at sea, including a fifth of the soldiers in the queen's navy. It killed 123 Londoners, flattened 300,000 trees, destroyed 900 homes and 400 windmills, and blew away countless church steeples, turrets, and lead roofs, including the one atop Westminster Abbey. Fifteen thousand sheep drowned in the Severn River on the storm surge at high tide.

More than recounting, Defoe tried to explain the storm to his read-ers. Emerging atmospheric science shows up alongside moral reflection and scripture, barometric readings hand-in-hand with metaphysical rumination. "I cannot doubt but the Atheist's hard'ned Soul Trembl'd a little as well as his House, and he felt some Nature asking him some little Questions," Defoe wrote. "Am I not mistaken? Certainly there is some such thing as a God—What can all this be? What is the Matter in the World?"

Searching for answers and counseling his readers, Defoe had done more than crank out the first modern work of journalism. He had is-sued the first modern weather report.

The earliest-known recorded rain science comes from the ancient Greeks; some of them had begun rolling their eyes at the prevail-ing belief that rain was sent by the almighty Zeus. In the Aristophanes play *The Clouds*, Socrates tells the farmer Strepsiades that Zeus doesn't exist. Strepsiades protests: "No Zeus up aloft in the sky! Then, you first must explain, who it is sends the rain; or I really must think you are wrong."

Only the clouds can send rain, Socrates tells him:

Was there ever a shower seen to fall in an hour
when the sky was all cloudless and blue?
Yet on a fine day, when the clouds are away,
[Zeus] might send one according to you.

About a century later, Aristotle refined Greek ideas about rain in his scientific treatise *Meteorologica*. Historians of science note the discourse gets it wrong on almost everything we know about weather phenomena—except rainfall. Aristotle saw rain as part of a sun-driven cycle among air, land, and sea, a rhythm he called "river of Ocean." Soaking into the earth, rain produced a "wet exhalation," forming springs as well as rivers that ultimately carried it back to the sea.

By the fourth century B.C., people had figured out some of the appreciable benefits of measuring rainfall; the more you know about what rain's done in the past, the better you can predict what it will do in the future. The first written reference is from India, in Kautilya's *Arthashastra*: "In front of the storehouse, a bowl with its mouth as wide as an Aratni [that's 18 inches] shall be set up as a Varshanana [rain gauge]." In Palestine, a book recording four hundred years of Jewish life through the second century A.D. reports a year of detailed rainfall and soil-moisture data. But those are isolated instances. The systematic measure of rainfall—still a key to rain science even in the era of weather satellites—would take another thousand years.

The cylinders that catch rain in modern backyards emerged in Korea during the reign of King Sejong the Great, who ruled from 1418 to 1450. A Korean cultural hero to this day, Sejong put a premium on science—especially agricultural technologies to help coax more food from the drought-prone land. Sejong wanted every village in the country to report rainfall back to the crown, a chore that involved inspecting roots and soils for moisture after a storm. His son, the crown prince, is said to have come up with the tubular gauge. King Sejong sent one to every village. Korean historians have a running disagreement about whether Sejong actually used the data, or collected it as a shrewd political move to show he cared about the problems of agriculture.

European weather watchers were using cylindrical catches by the time of the Great 1703 Storm; Defoe reports rainfall amounts captured in observers' "tunnels." But it is no surprise that the first rain instruments developed in the East. Legal executions for witchcraft were still well under way in Europe, and scientists were routinely hauled before the Inquisition.

In Italy, Evangelista Torricelli, the young mathematician-physicist who figured out in 1643 that mercury would rise and drop in a glass tube along with air pressure, kept his experiments secret to all but a few trusted compatriots. He'd watched the Inquisition of Galileo by the Church, and had gossipy neighbors who suspected he was up to witchcraft. Torricelli died of a brief and vicious fever when he was thirty-nine years old, before he could perfect his apparatus. In 1663, Robert Boyle named it the barometer—soon the centerpiece in a home weather-prediction craze that spread across Europe and then to America.

Torricelli also gave science its first description of wind, and a lyrical explanation of our place in the atmosphere: *Noi viviamo sommersi nel fondo d'un pelago d'aria*—"We live submerged at the bottom of an ocean of air." But when it came to describing and talking about the rain, neither science nor letters could ever sum it up quite so tidily. Rain's chaotic nature made it among the hardest parts of the weather to measure—and even to name.

n *So Long, and Thanks for All the Fish,* the fourth book in Douglas Adams's *Hitchhiker's Guide to the Galaxy* series, a lorry driver named Rob McKenna is a rain god but doesn't know it. "All he knew was that his working days were miserable and he had a succession of lousy holidays," Adams writes. "All the clouds knew was that they loved him and wanted to be near him, to cherish him and to water him."

McKenna despises rain, but he gets to know it so well that he creates a scale of 231 different types of rainfall. Type 11 is "breezy droplets," Type 33 a "light pricking drizzle which made the roads slippery," Type 39 "heavy spotting." Sea storms fall between 192 and 213. Type 127 is "syncopated cab-drumming." His least-favorite is Type 17, a "dirty blatter, blattering against his windshield so hard that it didn't make much odds whether he had his wipers on or off."

With its hurly-burlies and nor'easters, rain's eccentric vocabulary connects to the soggy literary landscapes of Ireland and England: Jonathan Swift is credited with the earliest published version of "raining cats and dogs" in 1738, though an English dramatist named Richard Brome had his dialogue raining "Dogs and Polecats" a century before. Some lexicographers suggest that, during bleak times, heavy rains might well have sent the corpses of drowned dogs and cats down streets and gutters—inspiration for Swift's gruesome mock pastoral "A Description of a City Shower."

Cat-and-dog cloudbursts seem practically ordinary compared with "raining young cobblers" in Germany. It rains shoemakers' apprentices in Denmark, chair legs in Greece, ropes in France, pipe stems in the Netherlands, and wheelbarrows in the Czech Republic. The Welsh, who have more than two dozen words for rain, like to say that it's raining old women and walking sticks. Afrikaans-speakers have a version that rains old women with knobkerries (that would be clubs). The Polish, French, and Australians all have a twist on raining frogs; the Aussies sometimes call a hard rain a frog-strangler. Portuguese- and Spanish-speakers both might say it's raining jugs. Inexplicably the Portuguese also say it's raining toads' beards, and the Spanish: *está lloviendo hasta maridos*—it's even raining husbands! Probably not what the Weather Girls had in mind with their 1982 hit disco single, "It's Raining Men."

Around the British Isles, hard rain is commonly described as persisting, pissing, bucketing, lashing, sheeting, stotting, or coming down in stair rods. In Scotland, people might say a heavy rain is chuckin' it doon, teemin', skelpit, stoatin' aff, or bouncin' aff the streets; a soft one that hangs in the air is a smirr or haar. Light rains have a graceful language with their mizzles and drizzles. In Ireland, a persistent drizzle is known as a "soft day."

Linguists mapping dialect in the soaking American South collected more than 170 descriptions for rain, including a temperance rain, a tub soaker, a log mover, a lighterd knot floater, a milldam buster, and a potato bed soaker. My southern father seems to have a hundred ways to describe a rain that hasn't even arrived yet, when the sky is "trying to rain," "wants to rain," or is "fixin' to rain."

This rich depth of description makes it odd that during the Scientific Revolution, as the bearded men of meteorology decided how to measure, classify, and talk about the atmosphere, they came up with only scant definitions for rain—light, moderate, or heavy, sometimes throwing in a shower or drizzle. The clouds earned themselves a much more elegant lexicon, and so did the wind.

We owe the expressive global language of clouds to an amateur meteorologist in London named Luke Howard, who in 1802 proposed a Latin-based classification that he likened to reading expressions on a person's face: Clouds "are commonly as good visible indications of these causes (of rain and other weather) as is the countenance of the state of a person's mind or body."

As a teenager, Howard built a small meteorological station in his parents' garden with a rain gauge, thermometer, and inexpensive barometer. His mother called the gravel path to the gadgets "Luke's Walk." Twice a day, he trekked faithfully down to record rainfall, evaporation, air pressure, wind direction, and high and low temperatures in his slim pocket journals. His father wanted his son's head out of the clouds, and sent him away for apprenticeship with a chemist. Howard became a pharmacist by profession, but studied meteorology all of his life, writing prolifically on clouds and the climate of London.

Howard's proposed classification was similar to the Linnaean system being used in botany and zoology, taken from the Latin for easy adoption "by the learned of different nations." His three primary descriptions: *cirrus,* from the Latin word for "fiber" or "hair"; *cumulus,* from the Latin for "heap" or "pile"; and *stratus,* "layer" or "sheet." Howard also suggested intermediate cloud types, various blends of the three primary clouds. For the rain cloud, which he saw as a stormy combination of cirrus, cumulus, and stratus, he chose the Latin word for cloud: *nimbus.*

Nearly a century later, in 1896, the world's top meteorologists gathered in Paris to mark a ballyhooed "year of the cloud" and release their agreed-upon system of ten types, based around Howard's names. *The International Cloud Atlas* is still the official identification guide, although over time, meteorologists have made it a bit clunky; in 1932, they reclassified Howard's rain cloud as a *nimbostratus*.

On the 1896 list, the king of clouds—towering *cumulonimbus*—was listed number nine. This is why, when we feel the highest of high, we say that we are on Cloud Nine. As the British cloud enthusiast Gavin Pretor-Pinney* tells the story, scientists unfortunately rearranged the order in the second edition of the atlas, shifting mighty cumulonimbus to number ten. But the phrase "Cloud Nine" stuck.

Today, *International Cloud Atlas* descriptions are used worldwide by scientists and fourth grade teachers alike to describe the fair-weather cumulus; the thin, wispy cirrus that often indicate a change in the weather; the low, flat stratus moving in with a drizzle. It's the nature of science, and also human nature, to organize Earth in universally recognizable ways. Well before globalization, we ordered our nations on maps, our music in scales, our geographic coordinates in long and horizontal lines. Just three years after Howard proposed his cloud nomenclature, a British Royal Navy officer named Francis Beaufort devised a scale for wind speed to give sailors a common way to describe wind and its impact on the sails of ships. The Beaufort Scale, tweaked for modern vessels, is still familiar worldwide. It is brilliantly simple for communicating the complexity of wind, says my science writer friend Scott Huler, who became so obsessed with its poetic elegance that he wrote a book about the scale. Beaufort 0 is "calm," described this way: "smoke rises vertically." Beaufort 1 is "light breeze," one to three miles per hour: "Direction of wind shown by smoke but not by wind vanes." In Beaufort 12, "hurricane," "devastation occurs."

All of which leads to an elephant in the room—a great gray missing from humanity's ordering of winds, clouds, musical scales, and vodka proofs. Among so many rigorous classifications, we don't have the same sort of global language for rain. There is no poetic lexicon shared by

---

* Author of a lovely book called *The Cloudspotter's Guide,* Pretor-Pinney is also founder of the Cloud Appreciation Society: cloudappreciationsociety.org.

ship captains and children's sky-watcher charts, no standard measure for precipitation that falls into the dizzyingly different types of rain gauges used by scientists: the totalizer, the tipping-bucket gauges, the weighing gauges, with various configurations among those and even different ways of measuring what they catch. Inevitably, both rain's description and its measurement are vernacular, often even personal.

Almost every atmospheric scientist, meteorologist, and weather reporter I interviewed about rain was either obsessed with rainfall as a child—watching it spill from the skies, tracking rain's runoff, and observing how it puddled and soaked the ground—or had a vivid weather experience in childhood. The latter was often more thrilling than terrifying, as Lance Morrow once well described big weather, "a child's delight in dramatic disruption."

So it was with the man called "the father of British rainfall," George James Symons. G. J. Symons was born in London in 1838 and "while quite young commenced regular observations of the weather." Historians of science trace his rain obsession to the severe droughts of England in the 1850s; it must have been sorrowful for the young rain watcher to see the skies turn dry.

By the time he was twenty-one, Symons had begun to build an enthusiastic club of rain-gauge readers, and to publish their collective data. The first issue of what would become a lifelong labor of love appeared in 1860 and included rain reports from five hundred stations. The work drew the attention of Robert Fitzroy, who had been tapped by the British government five years before to establish the country's first weather bureau.

Fitzroy was a distinguished naval captain who commanded HMS *Beagle* on its famous five-year voyage with Charles Darwin. One hundred and fifty years after Defoe's harrowing narrative, storms still caught captains by deadly surprise. In 1859 the Royal Charter Storm wrecked the ship for which it was named and as many as two hundred others, drowning more than eight hundred sailors. The loss inspired Fitzroy to develop a system of calculations and a new term, "forecast," to help alert the public and ship captains to coming weather. Even if imperfect, he believed that sharing rainfall, barometric readings, and other data could help reduce the great numbers of shipwrecks. The idea was enormously controversial in Victorian times. As they tried to save lives by predicting storms, Fitzroy and other public meteorologists were ridiculed as the "government Zadkiel," a reference to Britain's most notorious astrologer.

Fitzroy hired Symons as an assistant in 1860. But the young man was so obsessed with his British Rainfall Organisation that Fitzroy felt it took away from his official duties. Symons lasted only three years with the government's meteorological office. Fitzroy did not last much longer. He committed suicide in 1865 by slashing his throat with a razor. Fellow scientists felt sure it was the pressure of trying to make accurate weather predictions in the face of constant criticism and ridicule. His suicide sharpened the view that forecasting was an immoral pseudo-science. It helped lead to a ban on any public forecasts in England for the next thirteen years, allegedly because of inaccuracies. Darker motives were also at work. For one, the large ship-salvage companies of Cornwall and Devon complained to Parliament that the forecasts were putting them out of business.

Symons went on to collect and publish his rain data privately, which may have been the wiser route given the birthing pains at the British Meteorological Office, known today as the Met. He placed newspaper ads for rainfall observers in every corner of the British Isles. He offered to buy the instruments and train observers "of both sexes, all ages, and all classes." By 1865, the British Rainfall Organisation had a thousand reporting stations, by 1876 two thousand, and by 1898 three thousand. The dedicated volunteers checked their gauges at 9 a.m. every day, logged inches of rainfall on charts supplied by Symons, and sent them in once or twice a year.

Symons also dug up every historical record he could find, appealing for ships' logs and old weather journals so that he could reconstruct past weather. He ultimately gathered some seven thousand sets of records that let him build a reliable picture of rainfall all the way back to 1815.

With kind, crinkly eyes and a full Victorian beard, Symons was "a man of singularly genial manner making a friend of almost everyone with whom he came in contact, even those with whom he differed," a fellow meteorologist said of him. He maintained a patient and masterful correspondence with his rain army, sometimes cajoling, reprimanding, or polling the observers on various weather questions. He painstakingly recorded their thousands of annual readings, with town, observer's name, elevation, and total inches, in his annual publication *British Rainfall*.

The loyal rain-gauge readers in turn helped fund the organization with donations and subscriptions to his magazines. In addition to the measurements, *British Rainfall* carried anomalies and records from the year and exhaustive explanations of how to measure rain—on different terrains, for light rains, snows, and heavy rains, and how to note the total in decimal points by hundredths of an inch: "Vulgar fractions should never be employed."

Another publication, *Symons's Monthly Meteorological Magazine*, covered an astonishing variety of phenomena. The first volume, in 1866, ran features on a "new enemy to rain gauges"—the leaf-cutter bee; meteor showers; the migration of swallows; various floods and droughts; and an investigation into black rains, which we will encounter later.

By 1870, Fitzroy's successors realized what the Met had lost in Symons and his rain obsession. But Symons rejected offers to bring his network back to the public sphere. He told his observers their important rain work would not be pushed into "an obscure corner in some Government office." Such a move would undermine his own "enormous expenditure of money, time and physical and mental energy," and moreover mean the rain group's "esprit du corps would be extinguished."

When Symons died in 1900, he left no survivors, having lost his father when he was a young boy, then his mother, his only child in infancy, and his wife fifteen years before. But he was mourned by a family of thousands of fellow rain watchers. By all accounts he died a happy man, for having devoted his life to British rainfall.

Americans did not have the same qualms as the Victorian Brits about the voodoo of weather forecasting. Thomas Jefferson dreamed of setting up stations of weather instruments and deputies around the nation, but he was foiled by the Revolutionary War. Later, the telegraph made scientific forecasting tangible. Then and now, the best way to know how the weather is going to behave at your place is to ask its most recent host. In 1847—three years after Samuel F. B. Morse electrically transmitted his famous message, "What hath God wrought?" from Washington to Baltimore—America's first government meteorologist, the "Storm King," James Pollard Espy, pitched the idea of a national weather network connected by telegraph lines. Espy sold the notion to Joseph Henry, director of the brand-new Smithsonian Institution, who sold it to his board by stressing that telegraphing weather from the far western and southern reaches of the nation would "furnish a ready means of warning the more northern and eastern observers to be on the watch for the first appearance of an advancing storm."

By 1860, five hundred stations across the United States telegraphed weather reports to Washington. War, again, set back the effort; the network crumbled after the secession of the southern states. At the Cincinnati Observatory in Ohio, a young astronomer named Cleveland Abbe became so frustrated by the lack of public storm and flood warnings that he took on forecasting as a personal mission. With funding from the Cincinnati Chamber of Commerce, Abbe developed a system of telegraphic weather reports, daily weather maps, and predictions he compiled and shared via the Western Union Telegraph Company.

Western Union supplied outline maps, to which Abbe or a telegraph clerk could add symbols showing wind direction, areas of high and low pressure, and other details. Abbe designed the familiar triangular arrows and codes on hand-drawn forecast maps, with the "R" for rain. For the first time, Americans could read not only current weather conditions, but "probabilities," as Abbe called his forecasts, for the days to come.

Abbe's service was immediately popular. Only thirty years old, he became known as "Old Probabilities" or "Old Prob." It also became clear that forecasting could save many lives and ships. A petition from the Great Lakes region—which suffered 1,914 shipwrecks in 1869 alone—urged Congress to establish a meteorology agency and national telegraphic weather service to track "the origin and progress of these great storms."

Congress approved, and in early 1870, President Ulysses S. Grant signed a resolution that the secretary of war should head up meteorological observations in military stations across the nation, as well as "notice on the northern lakes and seacoasts, by magnetic telegraph and marine signals, of the approach and force of storms." Gen. Albert J. Myer, the army's chief signal officer and a personal friend of Grant, had pulled strings to secure weather in the War Department. Now, instead of shrinking in peacetime, the Signal Office could grow. Myer turned out to be as dedicated to forecasting as to devising the signal system. He quickly hired Abbe away from the Cincinnati Observatory. Abbe cranked out his daily *Probabilities* and trained a generation of meteorologists in the art of forecasting. Myer's signaling, developed for war, now grew into a system of flags that flew in cities and seaports to warn of coming storms. When he died in 1880, one newspaper tribute said "no careful seaman ventures out of port when the red light is burning or the red flag flying."

To this day, if you see a red flag flying with a black square in the center, be warned of the storm to come—and say a little thanks to General Myer and the U.S. Signal Service.

n a storm-gray federal government building in downtown Asheville, North Carolina, 160 scientists and other employees of NOAA's National Climatic Data Center work with weather records as new as satellite images transmitted a minute ago, and as old as the 1740s. The 20 million pages of observations, many handwritten, and billions more computerized records here represent the largest archive of weather data anywhere in the world. Logs of ship captains and military officers reporting in from around the globe are filed twenty-five feet high in the basement. "Your faithful servant," many of them sign off, in elegant, rounded script.

"A dash of rain" is an oft-repeated entry from the nineteenth century, the measure then as common as our dash of salt.

"The rain was driven with the force of arrows into my face, and the oppression was similar to what one feels riding on a fast horse at a riding pace," reads a harsh 1831 observation by the naval engineer William Redfield of New York, who subjected himself to storms as he worked to figure out the circular nature of cyclones.

If there is a common link in rain science past and present, it is a certain mystification. For countless reasons beyond the storm warnings that save lives, we humans have an insatiable will to divine rain—to manage the water supply of major cities, to figure out when to plant winter wheat, to plan an outdoor concert or choose a wedding date. But even as modern meteorology improves upon the old probabilities, rain is eminently difficult to predict or to count on. It's the classic example of chaos theory; as pontificated by Jeff Goldblum to Laura Dern in *Jurassic Park*: "A butterfly can flap its wings in Peking, and in Central Park you get rain instead of sunshine." The real scientist behind the butterfly theory was the late MIT meteorologist Edward Lorenz, the first to recognize that tiny, faraway triggers can change the weather in ways that make it impossible for mathematical models to predict. Your weather app can't tell you that it's about to rain old women and walking sticks in front of your house—while your backyard remains completely dry. It may not have predicted any rain at all.

Meteorologist Scott Stephens, a child cloud-watcher who knew his life's work by the time he was five years old, is charged with answering the National Climatic Data Center's public requests for weather knowledge past and predicted. Beyond the billion hits on the agency's website, he fields calls from police detectives and insurance investigators who need to know how much rain fell at the hour of a crime or accident scene; dam engineers seeking annual averages; construction bosses deciding whether to hire a crew; and the occasional crank who has figured out how to control hurricanes. (We'll get to that a little later.) These and the ten thousand other interested parties who call the center every year could find the historical data and predictions online, but they want the human insight. "Weather models do a great job forecasting temperatures, they do a great job of forecasting wind speed and direction," Stephens tells me. "The models have a more difficult time with precipitation."

This is true even in the Big Data age of supercomputers that crunch billions of global weather readings every day—from Earth-orbiting satellites; radars; ground sensors at thousands of stations; thousands more buoys and ships at sea; aircraft weather chasers; and one thousand beige weather balloons with white boxes known as radiosondes attached, launched every morning and afternoon from points around the world. (If ever you stumble upon a large, deflated balloon tethered to a plain white box, possibly making disconcerting noises and giving off a sulfuric stink, don't be alarmed, and return to sender in the postage-paid mailer inside.)

In the late 1940s, the military gave the Weather Bureau what became one of its most valuable tools for predicting rainfall—25 surplus radars, launching the age of weather surveillance. That network has grown to include 155 Doppler radars that let meteorologists peer into the clouds, and LIDAR, remote-sensing laser beams that can simulate floods. The United States and Japan jointly operate a satellite mission dedicated to measuring global rainfall. All of this data pours into NOAA's superspeed weather-prediction computers, which can crunch trillions of calculations a second, making forecasting more accurate all the time. Today's four-day rain outlook is as accurate as the one-day forecast of thirty years ago. The satellites and supercomputers have particularly sharpened forecasts for tropical storms; the National Hurricane Center nailed Hurricane Sandy's southern New Jersey landfall five days out.

Yet rain continues to defy Big Data—routinely washing out zero-percent predictions as well as the pronouncement by the former *Wired* magazine editor Chris Anderson of "a world where massive amounts of data and applied mathematics replace every other tool that might be brought to bear." What's fascinating about high-tech rain forecasts is the degree to which they are improved by human meteorologists like Stephens. National Weather Service statistics show meteorologists improve rain-forecast accuracy 25 percent over computer guidance alone.

Just as computer rain forecasts still need the instincts of human meteorologists, they benefit from the physical collection of rain—measured in gauges around the world—over radar and satellite imagery alone. For the Colorado state climatologist Nolan Doesken, this truth hit achingly close to home nearly twenty years ago. The National Weather Service's Nexrad (next-generation Doppler weather radar) failed to detect freakish rainfall variation in a storm that developed directly over Fort Collins, where Doesken lives. On July 28, 1997, many residents were lulled to sleep by the sound of rain drumming on rooftops, with no warning of what was to come. Before midnight, the storm poured two inches of rain over much of the region, but hurtled fourteen inches onto southwest Fort Collins. The rains swelled docile Spring Creek to a flash flood that swamped homes and sent families scrambling up trees and rooftops in their pajamas. Five people were washed to their deaths. Doesken has never forgiven himself for not making a call that night to the Weather Service; he never imagined the radar was not picking up such fierce rains.

The disaster led Doesken to launch a network of volunteer rain-fall observers, known as the Community Collaborative Rain, Hail and Snow Network, who take rain gauge readings at home and report them over the Internet. What began as a small local project has grown to 30,000 weather watchers around the United States who are building a database of highly local precipitation measurement. (About half of them consistently check their gauges every morning.) The physical data has proven invaluable for scientists, especially when it comes to rain's variability. A Texas volunteer once reported seven inches of rain in his Comal County gauge, while nearby volunteers reported no rain or a couple hundredths of an inch. Meteorologists were sure it was a typo until they looped back through the radar to find a tiny, short-lived convective storm that had formed and died in the same spot, circling the dedicated rain volunteer like the rain clouds chasing lorry driver Rob McKenna.

In April 2014, a storm carrying more rain than had hit the re-gion in any hurricane of the past century swamped Florida's Pan-handle, turning parts of Pensacola into sea and eating away chunks of the scenic coastal highway. The storm knocked out power to the official weather station at the Pensacola airport, but a number of the network's volunteers corroborated the extreme rainfall—reporting as much as 20 inches in one day—allowing scientists to confirm a new record. "The high-tech stuff will increasingly win the day," Doesken told me, "but it will win it better if it's ground-truthed" with credible measurement of rain falling in the backyards of tens of thousands of citizen scientists.

Doesken is a modern-day G. J. Symons (with salt-and-pepper mus-tache rather than Victorian beard), corresponding with his thousands of volunteers and meeting them in person when he travels. Most are older than fifty-five and so committed to the work that their fami-lies contact Doesken when they pass away. He writes back letters of condolence, not a duty he expected when he became a meteorologist. "These volunteers don't realize just how important their records can be," Doesken says. "Most of them find it interesting, but they don't realize that it can also be lifesaving."

Because we so urgently need to know what they know—every day, or even every hour—meteorologists have had to share their knowledge more publicly than most experts. From the earliest days of Fitzroy's forecasts in Britain and Abbe's *Probabilities* in America, a more parasitic group of weather watchers stuck close to the meteorologists like barnacles. Journalists were eager for weather news, the number one interest of most of their newspaper readers then; readers, listeners, viewers, and Internet users now.

My colleague Bill Kovarik, a professor of journalism and environmental history at Unity College in Maine, says early American journalists devoted ink to the weather long before the U.S. Weather Bureau existed. The *Niles' Weekly Register*, the most popular publication in the nation before the *New York Times* debuted in 1851, and then the *Times*, were both committed to weather reporting in the nineteenth century. *Niles' Weekly* ran lengthy reports on the atmosphere in its science section; in 1849, the magazine published an analysis of national rainfall patterns atop an update on aerial navigation and another on "the importance of fresh air." The *Times* was covering weather regularly by 1857, and had a daily weather report by the 1870s.

The addition of a newspaper weather map was a coup for newsmen who wanted to hook their readers, and weathermen who wanted their forecasts to reach a larger audience. With the help of the telegraph, meteorologists began making national weather maps in 1848; Joseph Henry hung one up for visitors each morning on the wall of the Smithsonian. In 1875 *The Times* of London became the first paper to print one daily. The *New York Times* began running one in 1934, and the next year, the Associated Press started to transmit a national map to its member papers. The early newspaper weather maps contained much more meteorological science than today's: isotherms—the lines that connect points with the same temperature—to help readers see where weather fronts were moving in; Abbe's little arrows to show direction; areas of high and low pressure. Like most things in newspapers, the maps were dumbed down steadily over the century, until they carried little besides temperature—and of course, the rain.

By 1900, the U.S. Weather Bureau had moved from the War Department to the Department of Agriculture, which helped set up the first radio weather broadcasts at the University of Wisconsin in Madison. Most fledgling radio stations shuttered during World War I, but UW's stayed on air, broadcasting weather news to ships sailing on the Great Lakes. During the New Deal, the Weather Bureau got much more involved with local radio, beginning a long history of love-hate relationships between the agency's meteorologists and local broadcasters who shared or hyped their forecasts.

The archetypal weatherman with his authority and verve was born not of TV, but of radio. One of the classics was Jimmie Fidler, still a student at Ball State University when he became "Radio's Original Weatherman" at WLBC in Muncie, Indiana. His show opened like this:

> *By telephone, telegraph, teletype, radio and the mail, WLBC's own meteorologist, Jimmie Fidler, "Radio's original weatherman," gathers the information on the weather as it is and as it is to be. Now, here is Jimmie with his maze of weather data that he will unravel into a simple and complete picture of the weather.*

When a handful of experimental television stations began broadcasting in the early 1940s, Fidler signed on with Cincinnati's WNBT, where he maintained his persona—trusty purveyor of "authentic weather information"—as TV's first weatherman. Other early approaches to TV weather hinted at the oddball art to come. New York City's first television weathercast debuted on October 14, 1941, and starred an animated sheep. Wooly Lamb, sponsored by Botany Wrinkle-Proof Ties, introduced each segment with a song: "It's hot, it's cold. It's rain. It's fair. It's all mixed up together. But I, as Botany's Wooly Lamb, predict tomorrow's weather." The meteorologist Robert Henson, author of several books about weather and two on the history of broadcast meteorology, calls Wooly the "harbinger of weather's eventual segregation from other television news."

Wooly remained on air for an astonishing seven years, but the early days of TV weather forecasting were mostly somber and serious. Many of the first broadcasters were World War II veterans who parlayed their meteorological skills into jobs in the nascent TV field.

The Federal Communications Commission inadvertently encouraged cheeseball TV weather in 1952 when it opened up competition for local station licenses. Most major cities went from one station to two or three. News managers vying for audience share found weather was the easiest news to liven up. "The result was TV weather's wildest and most uninhibited period," Henson writes, "the age of puppets, costumes, and 'weathergirls.'"

The Nashville poet-forecaster Bill Williams gave the weather in verse: "Rain today and rain tonight / Tomorrow still more rain in sight." In St. Louis, a puppet "weather lion" gave the nightly forecast. In New York, a sleepy bombshell in a short nightie gave the midnight forecast as she tucked herself into bed.

Weary of the indignities of fatuous forecasting, the American Meteorological Society tried to rein it in with a system of credentialing. "Many TV 'weathermen' make a caricature of what is essentially a serious and scientific occupation," the physics professor and Philadelphia weathercaster Francis Davis complained in a 1955 *TV Guide* piece, "Weather Is No Laughing Matter."

David Letterman didn't get the memo. Broadcasting the weather at his hometown station WLWI in Indianapolis out of college, Letterman "joked about 'hailstones the size of canned hams,'" Henson writes, "cited statistics for made-up cities, and once congratulated a tropical storm for reaching hurricane status."

Numerous stations hired meteorologists in the 1960s, but buffoonery—or beauty—continued to trump credentials. Before she became a movie star and sex symbol, Raquel Welch got her start doing morning weather in San Diego as Raquel Tejada, KFMB's "Sun-Up Weather Girl." The accomplished *ABC World News* anchor Diane Sawyer landed her first job out of Wellesley in 1967 as "weathergirl" for her hometown TV station, Louisville, Kentucky's WLKY. Besides her lack of meteorology experience, Sawyer recounts how she wasn't allowed to wear her glasses on camera despite terrible eyesight. She couldn't tell whether she was pointing to the West or East Coast on the weather map.

Meteorologists would put up with many more years of gimmicks before weathercasting became serious again. On NBC's *Today* show, Willard Scott delivered the weather dressed up as Carmen Miranda. "A trained gorilla could do this job every night," Scott once said of the forecaster's job. A very large gorilla, indeed, was about to change the profession.

With a wide smile and an even wider tie, John Coleman was the consummate 1970s weatherman, combining serious tweed-jacket forecasting with showmanship. *Good Morning America* launched in 1975 with Coleman as weather anchor. Breakfast tables tuned to him for seven years, and then his weather-broadcast dream came true.

Coleman believed the short time devoted to weather on TV— typically fifteen minutes a day—was inadequate for what viewers wanted and modern life demanded. He dreamed of a twenty-four-hour national cable network devoted exclusively to the weather. He spent every moment away from his day job developing the concept, figuring out how to program and staff an all-weather channel, and flying around the nation in search of a deep-pocketed financial partner. The partner would turn out to be Frank Batten, who had inherited a regional news-paper company, Landmark Communications, and turned it into one of the largest privately held media conglomerates in the nation.

Venture capitalists were skeptical of Coleman's plan. Batten could see what they could not. That's because he had been gobsmacked by weather since age six, when he and his uncle rode out a ferocious Category 4 storm, the Chesapeake-Potomac Hurricane of 1933, in the family's oceanfront cottage on Virginia Beach.

Coleman had many ingenious innovations. One was the proprietary technology to fit local forecasts and weather alerts into national programming. He insisted that all Weather Channel forecasters be trained meteorologists. And he convinced the National Weather Service to become the prominent source of information for the channel. For years, federal meteorologists felt bitter as the on-air TV personalities earned fame using their forecasts and data without attribution. TV broadcasters also might fail to report a warning or overhype one. The Weather Channel and National Weather Service struck a deal that got every warning issued by NWS out to viewers. It would be the greatest visibility government meteorologists ever had.

The Weather Channel launched on May 2, 1982, and rode out stormy years. The early technology garbled the local forecasts. Critics dismissed the channel as a joke. Coleman was a genius weather broadcaster but did not impress as CEO. Landmark invested $32 million to start the channel, and through '82 and '83 lost $850,000 a month. The company tried to push Coleman out, he fought back, and a rough legal battled resulted in a settlement that felt like a lose-lose.

By summer 1983, the board was preparing to shut down the channel. Viewership was so small that Nielsen barely registered it. Even if they didn't sit around watching it, though, Americans liked having the Weather Channel, and the nation's cable operators knew it. In the end, the operators saved the channel by agreeing to subscriber fees to keep it afloat. Starting in 1984, the fees coincided with the huge wave of growth in cable TV through the mid-1990s. The channel also found revenue in cheesy infomercials: "Heat-wave alert" for Ga-torade, "Cold-wave alert" for Quaker Oats, and "Weather and Your Health"—sponsored with no apparent irony by the fake bacon condiment Bac-Os.

The Weather Channel broke even after five years but continued to struggle with tiny ratings. The lack of pizzazz became obvious only in hindsight. Coleman had a strict "no remote feeds" policy—no live broadcasts from the field—because the technology was notoriously poor and expensive. The idea was for the forecasters to stay inside—and give viewers the weather information they needed to go outside.

As video equipment became better and cheaper, the channel's meteorologists began flipping this formula: They got out in the rain, while viewers stayed dry on their living room couches. It proved an incredibly appealing role reversal. Reporting from the field was a "sea change in our understanding of the emotional connection" people have with weather, said then-CEO Deborah Wilson. The macho meteorologist Jim Cantore cherishes his personal turning point. He'd been stuck behind a desk at the Weather Channel ever since he graduated from college in 1986, talking in front of weather maps. As Hurricane Andrew, Category 5, bore down in August 1992, a producer stretched thin for live feeds with other staffers in Miami asked Cantore if he'd like to pack a bag and cover Andrew's second landing in Louisiana.

Reporting on the weakened hurricane from his Baton Rouge hotel with the window smashed and the rain gusting in, Cantore expressed a love for storm drama that infected viewers. "It was awesome, the wind and the rain," Cantore says. "It wasn't a huge impact for Louisiana as it was for Florida, but it was awesome. It was my first one."

The Weather Channel streamed into the homes of 50 million Americans during Andrew; soon it was 100 million. In 2008, NBC and a pair of private-equity firms bought the channel for $3.5 billion. The man with the twenty-four-hour weather dreams didn't see a penny.

Coleman's settlement with the company had required him to turn over his 75,000 shares of stock, and since the Weather Channel was insolvent at the time, he received nothing.

Coleman landed as the local weatherman at KUSI-TV in San Diego, where his career took a surprising turn for someone who had devoted his life to explaining the atmosphere. In his seventies, he became an outspoken skeptic of human-caused climate change, using his platform to debunk science on the air, in local speeches, and on Fox News. He called global warming "the greatest scam in history." He joined an estimated quarter of television meteorologists, who, faced with the difficulty of forecasting tomorrow's weather, disputed the ability of climate scientists to make predictions fifty and a hundred years out. *Wheel of Fortune*'s host Pat Sajak, who began his television career as TV weatherman for KNBC-TV in Los Angeles, also turned in old age to spinning doubt about climate change.

Of course, climate scientists don't claim to predict the daily rainfall in fifty years, but the change in climatic conditions, a distinction that most meteorologists respect. Scientists and writers have come up with many good metaphors for the difference between climate and weather, such as the idea that climate is all the clothes in our closet, weather the outfit that we wear on a particular day. Or climate is what you expect—weather is what you get. (This and many other weather witticisms are often falsely attributed to Mark Twain. While Twain had many clever thoughts on weather, he did not conceive every atmospheric aphorism of human history.) My favorite analogy for climate and weather sees climate as the personality and weather as the mood. So weather is the mood of the atmosphere on any given day, in a specific place. Climate is the atmosphere's true personality—the average of these weather moods over many years.

I looked up Alan Sealls, chief meteorologist at WKRG in Mobile, Alabama, so that I could interview a weather broadcaster in the rainiest metro area in the United States. We made arrangements to chat about what it's like to forecast in a city with 100 percent chance of rain all summer long. But global warming and its impact on rainfall had become such a big story that we ended up talking mostly about climate change. In 2009, Sealls won a science-reporting award from the American Meteorological Society for his extensive series, "The Truth About Global Warming," during a time when most of his colleagues around the nation considered the topic "Kryptonite," in the words of one. Sealls has become a trusted voice on climate change and the human influence—even in a skeptical state. He stands at the crossroads of another profound shift in the profession. He is the best of the old weathermen, with a huge smile and personality to match; the best of the broadcast scientists, with a master's in meteorology; and the best of a new kind of weather watcher, helping the public understand one of the most complicated and urgent stories of our time.

He talks to his viewers about the long-term climate—at the same time he advises them to pack a raincoat today.

FIVE

THE ARTICLES OF RAIN

I n his 1615 memoir of the native people of Mexico, the historian Fray Juan de Torquemada described an ingenious local skill, one his men wished they could take home to Spain. The natives knew how to make rainproof garments. They tapped a milky sap from tall trees growing in the southern jungles, brewed the liquid over a fire, dipped in their capes and other clothes, then hung them to dry to a protective stiffness. They also plunged their feet in a batch to create a mold, then peeled it off, let it harden, and dipped it again and again. The sap shoes were waterproof, tailor-fit, and sturdy as mukluks.

For more than a century before Torquemada, Columbus and other explorers reported delightful bouncing balls made from the same goo, tossed for sport across Mesoamerica and South America. A century after, in a 1736 expedition to try to measure the curvature of Earth, a Frenchman and friend of Voltaire, Charles de la Condamine, sent a package of the sticky sap from Ecuador to the Paris Academy of Science. His exhaustive report included the native name: *caoutchouc*, from the Indian *caa ouchu*, "the tree that weeps." Condamine called the sap *latex*, the Latin word for a liquid or fluid. He carefully described the natives' method for boiling and smoking it into a pliable solid. But most of the scientists who bothered to examine the stuff found it a mere curiosity, useful only for rubbing out pencil marks—which gave it the name that would stick: rubber.

Condamine's Earth-measuring partner, the French botanist François Fresneau, was convinced the heart of the caoutchouc tree grew a miracle for man. He stayed on in South America to learn all he could about rubber, and devoted his career to figuring out its industrial uses. But neither he nor some of the best scientific minds of the eighteenth century could replicate the waterproofing techniques of the natives, whose patient boiling and smoking, Fresneau wrote to Condamine, "can only be executed on the spots where the tree grows, as these juices soon lose their fluidity."

Given the devastating rains of the Little Ice Age, the notorious storms of the North Sea, and the cultural fixation on rain in the British Isles, it's hard to believe westerners were well into the nineteenth century before they figured out how to rainproof themselves: rubberizing coats, cloaks, and carriage tops, and opening up umbrellas. Even then, it was all a remarkably tough sell. Shortly before he died, Fresneau figured out rubber was soluble in turpentine. But the early attempts to use it for waterproofing proved disastrous. Materials coated with the unwieldy brew melted in hot temperatures and cracked in the cold. Manufacturers didn't want to have anything to do with such products, which also smelled . . . well, like burning rubber.

By 1800, the single clever use of rubber was still rubbing out pencil.

One day in 1819, a British inventor named Thomas Hancock was cursing a batch of rubber and turpentine as he tried to fashion waterproof roofs for horse-drawn carriages. He botched the batch, a fortunate mistake. He ended up with elastic, which he patented the following year, for attachment "to the wrists of gloves; to waistcoat backs and waist-bands; to pockets, to prevent their being picked."

Hancock had a mind for mechanics but little aptitude for chemistry. Hard as he tried and close as he got, he could not quite figure out how to make waterproof fabric, which he desperately wanted to manufacture the beautiful rainproof "articles" he'd sketched in his journals: coats and cloaks, sea hoods and fishing boots. Solving the puzzle would take someone with a preternatural grasp of chemicals. While Hancock was breathing turpentine fumes in London, an accomplished chemist in Scotland was a batch closer to the invention that would make his name synonymous with the raincoat—far beyond his imagination, his continent, and his time.

Charles Macintosh was born in 1766 in Glasgow, where his father was a prosperous merchant and dye maker. The hardscrabble tobacco port was just starting its climb as Scotland's industrial powerhouse. Glasgow was the major European entry point for raw American tobacco until the colonies won their independence and the trade collapsed, taking the burgh's economy with it. Cotton, coal, and chemicals all helped restore the tobacco wealth and more. Water-powered cotton mills were rising on the gritty banks of the River Clyde; in 1795 Glasgow had a dozen and by 1839 there were nearly one hundred.

Young Macintosh went away to school in England and returned home for apprenticeship in a counting house. He was supposed to learn mercantile affairs and help sell his father's goods, but his mind was captivated by chemistry. The new scientific discipline was just coming into its own, out of the miasma and superstition of alchemy. Macintosh had an ether-clear native talent for mixing and morphing its elements. At eighteen, he was corresponding with the well-known chemists—most then were physicians—of Scotland and England, inquiring about chemistry lectures and how he might make colors from vegetables. He began traveling to Edinburgh to study with Joseph Black, a medical professor who had discovered "fixed air," soon called carbon dioxide. Black, with Macintosh and some other students, formed the earliest-known chemical society. Before Macintosh turned twenty he had written society papers on alcohol, alum, crystallization, and "the application of the blue colouring matter of vegetable bodies."

He was not yet twenty-one when he quit the counting house to set up his own plant to manufacture sal ammoniac, a crystalline salt in high demand to make everything from tinned copper to pharmacy cures. Macintosh had secret sources for his salt: he extracted it from soot and urine. Human waste was easy to come by in the city now becoming crushed with poor immigrants fleeing potato famine in the Highlands or evicted from their farms in Scotland's forced displacements known as the Clearances. For many years, Macintosh's father had paid for pee. The poor would save up the family's urine and hand it over to landlords when it came time for pickup by George Macintosh's collectors. The elder Macintosh used the ammonia to manufacture cudbear, a coveted red-purple dye made from lichens.

Thriving on Scotland's rain and mist, the country's abundant lichens are central to its textile history; the fawn colors (and slightly funky smell) of Harris tweed come from lichens in the *Parmelia* genus. Most cudbear manufacturers used a Scottish lichen called *Ochrolechia tartarea*, but George Macintosh imported more-exotic types from the Italian island of Sardinia. In his twenties, Charles Macintosh traveled across Europe for months at a time to scout lichens, flowers, and plants for potential new colors and materials, or to meet with possible business partners. His surviving papers don't indicate how long he had pondered waterproof cloth in the years or decades before his famous brainstorm. Perhaps it crossed his mind as he walked the puddled cobblestone streets of Glasgow, or in the spring of 1789, when he experienced a terrifying storm in the passage between Sunderland on the east coast of England and Rotterdam in Holland. On that trip he visited the Kingdom of Prussia to try to land the contract for dyeing the Royal Prussian Army's uniforms blue. Macintosh always seemed more comfortable with the chemistry than the commerce. The Prussians turned him down. Surely he would have clinched the deal if he could have kept them dry as he perfected their blue hue.

In the ethos of his father's generation of dye makers who sourced ammonia from Glasgow's human urine stream, Macintosh had a nothing-wasted mind-set. His discovery of the long-sought solvent for rubber came out of his search for uses for some of the nastiest by-products of early nineteenth-century progress. Gas lamps were becoming popular in the cities of Europe, lighting up the wealthier streets and private homes. But the tar sludge left behind in the manufacture of coal gas was a public menace—growing in piles in the Thames in London and the Firth of Forth in Edinburgh. Macintosh saw practical uses in the sludge and wastewater, which included valuable ammonia. In 1819, Glasgow Gas Works was only too happy to sign a contract to sell him all the waste it produced; the company had been dumping it into rank pits around the city.

Macintosh converted the tar into pitch (at that time used to waterproof wooden boats and crates), and separated the ammonia for cudbear. That left him one more by-product. He suspected it might be useful, but it was also a humdinger. The pitch-making left behind a volatile liquid called naphtha. Highly flammable, naphtha put the fire in "Greek fire," a deadly chemical weapon from ancient times. Arrows dipped in the stuff could arc inextinguishable flames into a village. Shot through a brass cannon in a stream of fire, it could incinerate a line of soldiers or an entire ship. "Every man touched by it believed himself lost, every ship attacked with it was devoured by flames," wrote a crusader in 1248. Greek fire adhered to victims and kept burning in water, so even jumping into the sea would not quench the flames. Under Pope Innocent II in 1139, the Second Council of the Lateran had decreed it "too murderous" for use in Europe.

The decree was still honored by the time Macintosh began experimenting with the naphtha from Glasgow Gas. He thought it might be the one substance powerful enough to tame rubber into a waterproofing varnish. The winning recipe turned out to be ten to twelve ounces of shredded rubber combined with a "wine-gallon" of naphtha. Macintosh heated the brew as he ground it into a thin pulp, then ran it all through a fine sieve until it resembled "thin, transparent honey."

Macintosh had managed to make a waterproof brew, but it gave fabrics a clammy look and feel—and a smell sickening enough to turn a urine collector's stomach. Then he had an idea simple and brilliant as a sandwich. He spread his warm honey between two pieces of fabric, and pressed them together using heavy rollers. The resulting cloth was flexible and waterproof. In 1822, he obtained patent number 4,804, for "Waterproof Double Textures."

The first mac was born.

Macintosh was certain he could sell his textures to clothiers eager to bring waterproof articles to the masses. He set out to build a mill to manufacture his new invention. The only financing he could land required that he locate it two hundred miles south of Glasgow, in the textile behemoth of Manchester, England. His new business partners were a pair of brothers named Birley, owners of a large complex of cotton mills in Manchester's Chorlton-upon-Medlock district. Hugh Hornby Birley and his younger brother Joseph agreed to finance Macintosh's rubberized cloth factory. But they remained suspicious enough about the technology and the product that they wanted Macintosh's mill adjacent to their own. Should he fail, they could absorb the space.

At first, the Birleys' skepticism seemed on mark. Even locked between fabrics, the rubber and naphtha were never completely tamed. Like the earlier misfires, Macintosh's rubberized cloth sometimes melted in hot temperatures or stiffened in the cold. Naphtha's nauseating odor clung to finished products—like lichen on Harris tweed but much harder to live with. Rough and unrefined, the fabrics did not spark the interest of the fashionable—though a market began to build among the armed forces.

In a marketing coup in 1824, Macintosh outfitted the Royal Navy officer John Franklin for his third and most successful expedition to the Arctic. When the beloved explorer placed a large order for waterproof canvas, Macintosh threw in a waterproof pillow. Franklin wrote back requesting "four life preservers of a size for stout men, and eighteen bags about six feet long and three broad, fitted with corks for filling with air for the party to sleep on, and four for pillows of the size of the one you gave me."

Still, Macintosh was not able to sell his fabrics to the clothiers, even those making coats and cloaks. He needed someone who understood the potential for keeping Europeans dry in the rainy outdoors and at sea, and who could help the public see it. That would turn out to be Thomas Hancock, the British inventor with a mind for mechanics who'd been sketching his articles of rain.

In 1825, Macintosh agreed to license his double textures to Hancock to make the articles in his books, full of elegant pencil drawings of waterproof coats and trousers, boots and hoods, bathing caps, traveling cushions, hoses, and even tires—six decades before John Dunlop invented rubber tires for bicycles and filled them with air.

While Macintosh continued to fill orders from the likes of the navy, Hancock got to work trying to improve the fabric for his articles. One advantage he had over Macintosh was an invention he called the masticator. (In the early years of the invention, Hancock obliquely called it "the pickle" to prevent anyone from stealing it.) The machine heated leftover scraps of rubber as it chewed them up, making the material pliable without chemicals. At his large mill on Goswell Road in London, Hancock hitched horses to masticators and large iron rollers to fabricate rubberized sheets he sold in the shipping and yachting industries.

Once Hancock understood Macintosh's process, he came up with a blend for the cement—less naphtha and more turpentine—that made the rubber easier to handle and slightly better smelling. But it took years to convince Macintosh to take him on as a partner. The older man was wary, and kept his competitor at arm's length as they manufactured their waterproof goods separately in London and Manchester. It was not until Hancock's articles began outselling Macintosh's that the Scottish chemist brought his rival to Manchester. In 1831, he made Hancock a partner in Macintosh & Co.

Unlike Hancock, Macintosh never really wanted to make the articles themselves. He would have been happy to manage his considerable chemical concerns back in Glasgow and the rubberized fabric mill at Manchester. He became the world's most famous raincoat maker reluctantly, only after the clothiers refused to pick up the idea.

Macintosh and Hancock's first garments were not rain coats, but rain cloaks. Men and women had worn cloaks, capes, and ground-sweeping mantles since the first century A.D. In the early nineteenth century, the billowing twills were beginning to give way to great-coats—a combination of the cloak and today's calf-hanging coat. Yet cloaks and capes remained most popular for foul weather, often oiled to deflect rain.

The two inventors were heartsick to realize that brainstorming the waterproof fabric would be the easy part compared with the struggle to convince people to wear the rainproof cloaks. Physicians were in large part to blame. Some doctors were convinced, and convinced their patients, that although Macintosh & Co.'s cloaks kept out rain, "they stopped perspiration, and hence were injurious to health."

Hancock claimed the doctors had an ulterior motive: They secretly feared rain cloaks would make people healthier, cutting into their business. He also blamed the merchants and customers who fit the garments too closely, causing unnecessary perspiration. "Complaints arising from this source long annoyed us, and exposed us to no end of abuse."

Seams and buttonholes proved another nightmare. Each stitch acted like a tiny straw that sucked rain inside the coat and soaked its wearer.

Europeans' early disdain for swaddling themselves in waterproof cloth may have been equaled only by their reluctance to hold their own umbrellas. It was as if God didn't want them to spurn His heaven-sent creation. At least, not if there was a servant to do it for them.

In the early eighteenth century, only footmen and servants used umbrellas practically, keeping them ready by the doors of dining rooms and inns to escort clients to and from their carriages. Upper-crust women carried fancy parasols, a matter of haute couture rather than keeping dry. An umbrella in a man's hand was the ultimate sign of effeminacy. If he had to walk in the rain, a gentleman should wrap himself in a cloak or coat made of beaver felt, and cover his head with a beaver-felt Wellington—just a few of the naturally waterproof articles then driving the breakneck expansion of North America over one furry rodent.

The slave trader turned abolitionist John Newton, better known for writing "Amazing Grace," put the social stigma this way: "To carry an umbrella without any headgear places a fellow in a social no man's land in the category of one hurrying round to the corner shop for a bottle of stout on a rainy day at the behest of a nagging landlady."

Ultimately, the Brits had to pick up the umbrella, essential prop in the strut of the derby-hatted gentleman and a stealth weapon for the unarmed Sherlock Holmes. (Sir Arthur Conan Doyle gave him a mac, too.) The umbrella, open or closed, is, like the bicycle, one of human-kind's very few utterly functional inventions that also happen to be completely beautiful. In his charming book about U.K. weather, *Bring Me Sunshine,* the British author and broadcaster Charlie Connelly pro-fesses to love rain. But never does he write so lyrically as in his chap-ter devoted to the dignified umbrella, "with its smooth, symmetrical flowering as you put it up, the effortless movement and coordination of countless working parts, the elegance of its dome—the umbrella is a beautiful machine." Seeing one "battered and ruined and shoved into a bin" on the sidewalks of London always sinks Connelly's heart.

Connelly celebrates Jonas Hanway as the man who finally de-mocratized the brolly. The respected reformer governed a hospital for deserted children and worked on behalf of many other social causes. Those included fighting the widespread drinking of . . . tea, newly pop-ular in the London coffeehouses. No doubt shielding himself from the city's growing proliferation of tearooms, teahouses, and tea gardens,

Hanway defied eighteenth-century etiquette to become the first gentleman in London to carry an umbrella everywhere he went. Rain or shine, the article was his signature for thirty years. Hanway ignored the early stares and tongue-wagging. By the time he died in the fall of 1786, umbrellas were becoming must-haves, like the lampposts rising on the damp streets of London. Surely that weather watcher Daniel Defoe and his 1719 novel *Robinson Crusoe* also helped popularize— and defeminize—the umbrella. Defoe's shipwrecked castaway labors for weeks to make himself a sturdy goatskin umbrella. It's a gruesome article made of hide and hair, though later artists and book jackets often softened the image, depicting Crusoe's creation as a pleasant dome of leaves or palm fronds. Crusoe describes his umbrella as clumsy and ugly, but "the most necessary thing I had about me, next to my gun." It is one of the few memories of his island that he takes back to London upon his rescue. Londoners kept Crusoe's umbrella in their hearts, too: with the popularity of Defoe's novel, the British began to call umbrellas "robinsons."

One cloudless November day, I visited Europe's oldest brolly shop, James Smith & Sons Umbrella Shop, which opened in London in 1830. (Props to Hanway and Defoe.) Wooden and custom-order umbrellas and canes are still handcrafted in the basement. The street-level showroom feels like a magical emporium, just the place where Mary Poppins would have purchased her handled hovercraft. Umbrellas line every inch of wall, handles all facing forward like an army at the ready to battle London's biggest storm. Still more are gathered like bouquets in wicker baskets on the floor, displayed in glass cases, and hung in windows and racks, classic blacks, greens, and navies in one section, a rainbow of solid colors in another, patterns in another. Sticks and handles shine in solid maple, ash, hickory, and cherry, some decorated with a blue crystal parrot head or a hand-carved hound. A dapper staff of young men also stood at attention, I hoped not facing a lifetime of bad luck for how often they must open an umbrella indoors. I asked a bow-tied salesman my most urgent question: Are they ever asked to customize an umbrella to hide a secret weapon like in the movies? Or real life. The Bulgarian dissident writer Georgi Markov was assassinated by a brolly as he waited for a bus on London's Waterloo Bridge in 1978. He felt a sharp pain on the back of his thigh, and looked behind him to spot a man grabbing up an umbrella and jumping into a taxi. That night Markov developed a fever. He died three days later of poisoning from a ricin-filled pellet. The young salesman told me that James Smith & Sons declines to fashion umbrellas for weaponry. (An answer that led me to surmise they do get asked.)

Browsing all the useful articles and their brilliantly simple technology, it was hard to believe people once spurned umbrellas. But the Georgian-era Brits who would not carry their own were stuck in the muck of custom. The umbrella was the purview of servants from its earliest design, which seems to have been to protect from sun rather than rain. The first depictions come from hot, arid climates such as Mesopotamia. Just around the corner from James Smith & Sons, at the British Museum, a gypsum wall panel excavated from the palace of King Ashurnasirpal II shows the king clutching some arrows to declare a victory; over his head, a servant holds a parasol on a long stick. The relief is about 3,000 years old. The art writer (and umbrella aficionado) Julia Meech says runners visible in this and other depictions from Mesopotamia suggest the Assyrians invented the first collapsible umbrella. The earliest artifact of an actual umbrella also comes from this part of the world: a wooden top notch with eight socket fittings for ribs, found in an eighth-century Phrygian tomb at the ancient city of Gordion in Turkey.

In Egypt, the umbrella was associated with the vault of heaven—tied to both sun and rain. Nut, the mother goddess of ancient Egypt, evoked a gigantic parasol, her body sheltering all the Earth. Indian rulers more than 2,000 years ago received a white umbrella at their coronation, representing sovereign power over the world. For Buddhists and Hindus, the umbrella evolved as a symbol of comfort and respect: Remember Lord Krishna lifting Mount Govardhan over his sheepherder friends?

The umbrella was especially significant in China, where Confucian texts imagined the twenty-eight ribs of a chariot parasol as stars and the central stick as cosmic pillar and axis of the universe. Carried over Zhou kings, the umbrella signified omnipotence as well as practical protection from sun and rain. Ultimately, the Eastern world came to democratize the umbrella just like in the West. In fact, China's greatest gift to the articles of rain would turn out to be wonderfully humble.

The articles of rain are often tough and stout: clumping galoshes, yellow slickers, fishermen's hats ready for sea. Crinkling open as exquisite exceptions are the oiled paper and bamboo umbrellas of old Asia. So delicate, they would hardly seem matched for a storm.

The handcrafted umbrellas actually stood up beautifully to rain. They became particularly popular on the island of Honshu, where they are called *wagasa*. Prior to the 1950s, Japan's streets were filled with them. The artist Stephan Köhler described them as private skies, hovering over every age and class of person.

Like in the rest of the world, umbrellas in Japan's earlier history were exclusive articles for nobility and feudal lords, held over heads by hereditary umbrella bearers with strict rules about which colors could shade which rank. Most Japanese fended off rain with sedge hats tied under chin, water-repellent straw capes, and oiled paper raincoats called *kappa*. But in the late seventeenth century, more and more people in the largest cities began carrying the status-suggesting umbrellas. Around 1800, *wagasa* exploded as the national daily accessory—thin paper an ideal sunshade, waterproofed with lacquer and fortified with bamboo ribs to deflect even the strongest rains and winds.

Throughout Japan, *wagasa* generated a thriving craft economy in umbrella-making districts such as Gifu City, where up to sixteen craftspeople contributed to each one: hand-making the papers and dyes; carving the bamboo stick, intricate ribs, and other parts; stringing the mechanisms with silk thread; painting, oiling, and lacquering; adding special decorations for tea ceremonies or dance. At an industry height in 1950, Gifu City alone produced 15 million *wagasa* in a year.

In the late nineteenth century, strange imports with steel ribs began showing up in the hands of foreigners living in the open port of Yokohama. Samurai wanting to appear civilized took to carrying the lightweight Western umbrellas. Japanese people, curious and enthusiastic about the West, soon began to fold up their *wagasa*. After a short resurgence post–World War II, the industry sank into a steady decline. Today, only a few traditional umbrella makers remain, crafting a small number of *wagasa* for festivals and professional dancers. Like the kimono and the paper fan, the *wagasa* is now best remembered in art and dance.

I n the history of the great articles of rain, skepticism is the common thread. The clever windshield wiper is another example. But, unlike Charles Macintosh and his raincoat, the inventor was never credited for the rubber metronomes that allow us to see the road even in a frog-strangling rain. Meet Mary Anderson, a society belle from Birmingham, Alabama. Like Ada Lovelace, who described how computing machines could solve math problems, or the biophysicist Rosalind Franklin, who was the first to photograph the DNA double helix, you've probably never heard of her.

At the turn of the twentieth century, when automobiles were still a novelty, drivers rattled about in chaos—frequent breakdowns, no traffic lights to guide them, no filling stations to fuel the high-wheel motor buggies sputtering through cities in the days before Henry Ford's models for the masses. Driving was a misadventure in clear skies. Come rain or snow, it became misery mechanized. To clear what were then called windscreens of rain and fog, drivers would stretch far out their doors and wipe the screen with their hands. Sopping wet arms were the least of the consequences.

The blessed but slow end to manual windshield wiping began in 1903, after Anderson took a stormy trip to New York. She watched the male drivers making a holy mess and came up with a solution. That June, she applied for a patent: "My invention relates to an improvement in window-cleaning devices in which a radially swinging arm is actuated by a handle from inside a car vestibule," she wrote in her application. "A simple mechanism is provided for removing snow, rain and sleet from the glass in front of the motorman."

Anderson's mechanism consisted of a rubber blade attached to a spring-loaded arm that would sweep across the glass as the driver cranked it from inside. She landed U.S. patent number 743,801. Automotive engineers scoffed. They thought the wiper would distract anyone who had to operate it while simultaneously watching it flap about. Anderson's patent expired before her idea was widely adopted. By 1916, most vehicles manufactured in the United States came with windscreen wipers as standard.

The following year, another woman entrepreneur, Charlotte Bridgwood, president of the Bridgwood Manufacturing Company in New York, was awarded a patent for the first automatic windshield wiper. Her electric Storm Windshield Cleaner relied on rollers rather than blades. Her patent, too, expired before Henry Ford made automatic wipers a standard feature on his cars just six years later—never acknowledging Bridgwood's contribution.

In the spirit of the robinson and the mackintosh, I hereby propose that we all start calling windshield wipers "marys," in honor of their original inventor, Mary Anderson.

Macintosh and Hancock eventually figured out how to rainproof their seams, by smearing them with rubberized glue, which got a little better-smelling all the time. People started wearing the cloaks and coats correctly. The company caught attention when it designed the Duke of York's military cloak—a waterproof blue lined with crimson silk. Likewise when army guards began to wear light drab capes by Macintosh & Co., other young men wanted to follow their example, "especially of the drab color," Hancock wrote.

Charles Macintosh lived to see his waterproof cloaks and coats "take with the public," as he always thought they should. He died in Glasgow in 1843. One year later, on May 21, 1844, Hancock was awarded the first patent for the vulcanization of rubber. Named after Vulcan, the Roman god of fire, the long-sought invention cured rubber by heating it with sulfur, making it as pliable as the European explorers had seen in the South American jungles hundreds of years before. The American inventor Charles Goodyear filed his U.S. patent for vulcanization just three weeks later, and claimed that Hancock had stolen it after a colleague of Goodyear's shared it with him in London. Goodyear lost a legal battle with Hancock, and died in 1860 with no assets. Hancock lived five years more, dying a wealthy man in London at the age of seventy-nine.

By the 1880s, "mackintosh" was the household name for a raincoat in Europe. When and why the "k" was added remains a mystery. Hancock never uses the misspelling in his journals, so it must have been introduced shortly after his death. The Victorian novelist Mary Augusta Ward reflected the common use in her 1888 bestseller *Robert Elsmere*. During a breathless deluge, a soaked beauty protests and then submits to wearing the protagonist's raincoat: "He put the mackintosh round her, thinking, bold man, as she turned her rosy rain-dewed face to him." By century's end, mackintoshes were popular in the United States as well. The 1897 Sears Roebuck & Co. catalog offered "Men's Double Texture Mackintosh Coats, with large full size cape and velvet collar."

Dunlop Rubber bought Macintosh & Co. in 1925 and continued to sell all the old articles, though its major focus was industrial rubber production and tires. Dunlop made mackintoshes (the "k" by now

official) for the British army during World War II, British Rail workers, and London's Metropolitan Police, which helped give the coat an
unlikely sex appeal. Playing tough gumshoes in the 1930s, Humphrey
Bogart also showed how dapper a man could look in a mac. When
Ingrid Bergman and Audrey Hepburn started wearing them, every
schoolgirl wanted one, too.

In 1953, Queen Elizabeth and her husband, Prince Philip, were
photographed in the pouring rain, he looking archaic in a royal cape
of fur, she elegant and modern in a tan-colored mac. Long before a
twenty-one-year-old McDonald's advertising secretary named Es-
ther Glickstein Rose named a new stacked hamburger, and Apple's Jef
Raskin code-named a secret computer project after his favorite apple,
the mac was a global brand.

I n the counterculture milieu of the 1960s, any articles popular in the
turn-of-the-century Sears catalog—suspenders, ladies' feather hats,
and rococo couches among them—became contemptuously old. The
mackintosh began to feel less classic, more old-fashioned. Dunlop,
now juggling products from tires to tennis rackets, had dropped the
ball on fashion branding. Mackintoshes were now utterly generic,
manufactured in hundreds of factories, only a few of which adhered
to Macintosh and Hancock's standards of hand-cut coats with rubberized seams. Cheap new plastics led to tacky "fakintoshes." Mackintoshes slowly earned a reputation as your grandfather's raincoat—or a
dirty old man's. The once-respectable coat somehow became associated
with flashers, movie-house masturbators, and other sexual deviants.
The British slang dictionary *Knickers in a Twist* carries an entry for
the "Dirty Mac": "a worn and besmirched raincoat synonymous with
perverts and gentlemen of the tabloid press." (Today, old mackintoshes
are a favorite article among rubber and raincoat fetishists—the best of
both worlds.)

Also during the '60s, the father-and-son chemical engineers Bill

Gore and Robert Gore had been experimenting with polytetrafluoro-ethylene (PTFE for short, a compound used in Teflon pans and circuit boards) in their Newark, Delaware, basement. They were cooking it up in pots, pans, and a pressure cooker when Robert figured out how to stretch the stuff to one thousand times its size. The material was porous but incredibly strong, chemically inert—and breathable. Their family-owned company went on to develop hundreds of medical, industrial, electrical, and textile products from the material, including a fabric they registered as Gore-Tex. While Robert Gore ran the multimillion-dollar firm, his adventuresome parents, Bill and Vieve, went on all manner of camping and hiking adventures to test out the light outdoor products that could stand up to rain. After much trial and flooded-tent error, the first Gore-Tex raingear hit the market in 1976.

It was the biggest revolution in rainwear since Charles Macintosh's Waterproof Double Textures. Except this time, clothing manufactur-ers and consumers raced—on runners' and hikers' legs, on bikes, and on skis—to outfit themselves in the new articles. Companies all over the world began introducing their own breathable waterproof lines of gear. Now mackintoshes were looking really fusty. It is perhaps no surprise that the man who would restore them in name and style was a Scotsman, born in Glasgow just like Charles Macintosh.

Growing up east of Glasgow in a post–World War II new town called Cumbernauld, young Daniel Dunko might have seemed an unlikely fashion scion. His father was a displaced Ukrainian who worked as a slabber after the war, building dams across the Highlands, where he met a Scottish girl who would become his wife. The Dunkos had four sons, and moved them to Cumbernauld so they could grow up with a garden and fresh air. Daniel was the spoiled youngest, "the boy allowed to have pudding before the main course."

Perhaps the indulgence was the entitlement he needed to succeed in fashion and bring back the mac. Dunko left school at sixteen for a training spot at a coat factory where his oldest brother worked. Traditional Weatherwear, based in Cumbernauld, was one of the few clothiers that still hand-cut and hand-glued mackintoshes from the old double textures. The company made raincoats for police departments and sewage workers throughout the U.K.—the latter in olive green to hide the stains of the job.

The raincoat maker's index finger is his most important tool. Plunging it into a pot of strong-smelling glue, he scoops up a blob, carefully smears it along the seams, then reinforces each with a strip of rubberized cotton. After three years as an apprentice, Dunko could not imagine a future confined to the tip of his pointer finger. He appealed to the company's owner to move him to sales, without success. After two more years, the owner finally relented, warning that should Dunko fail, there would be no tailor's job to return to. Dunko didn't know how true that was. The company was on the verge of bankruptcy.

Handsome, with the aquiline face of a young actor and long fingers more suited to spreading swatches than glue, Dunko turned out to be very, very good at selling articles. He pushed the coats to luxury houses like Gucci, Hermès, and Louis Vuitton. In 1996 he was named director of sales. Dunko was convinced the future lay in premium fashion; he wanted the company to give up utilitarian raingear and its dull reputation. Ultimately, the fusion of heritage with high fashion was the charm. Traditional Weatherwear trademarked the name "Mackintosh Made in Scotland," stitching a Scottish dandy on each label. The line took off. Dunko took out a $100,000 bank loan to acquire a 10 percent stake in the company. In 2000 he rounded up investors to buy it. In 2003, twenty years after joining Traditional Weatherwear as a teen apprentice, Dunko officially changed the name, to Mackintosh Rainwear.

The repurposed mac was a particular hit with the fashion-crazed Japanese. The coats were soon more popular in Japan than in Scotland. Dunko propelled the trend by launching Mackintosh retail stores in Tokyo, dressing in a kilt for each grand opening. Dunko tapped

Japanese retail mogul Yuzo Yagi as his importer. Yagi started to court Dunko with an eye toward making Mackintosh part of his retail empire, Yagi Tsusho. Dunko knew the company's import-export network would solidify Mackintosh as a global brand. He knew Yagi could restore the mac to its pre-pervert heyday, when the coats were worn proudly by the queen and Bogart. What he did not foresee was that Yagi could do all of that without him. After selling Yagi the company for £7.5 million in 2007, Dunko stayed on in Scotland as managing director—only to be edged out four years later amid clashes with the board.

When I visited Dunko, he had just opened a new raincoat factory in Cumbernauld. In the artfully sparse studio, his oldest brother oversees coat makers who hand-cut the rubberized fabrics in tennis green and honeysuckle red, pushing glue along the seams with muscled pointer fingers. In addition to coats, Dunko is making evocative new articles of raingear, including sea hoods and a waterproof high-top sneaker that's been picked up by Converse. The items were sketched by a young U.K. inventor two centuries ago. Dunko is building his new company around the memory of Thomas Hancock, and the dream of his articles. He named it Hancock Rainwear.

A few miles away, the Yagi-owned Mackintosh factory operates from the same building where Traditional Weatherwear made coats since 1972, boxed in by a frozen-food conglomerate. The morning I visited, the radio blasted the American pop single "Thrift Shop." Rich-colored bolts of fabric lined the walls of the high-ceilinged warehouse. The wools, flannels, cottons, silks, and cashmeres were all rubberized in Charles Macintosh's famous sandwich, but thin and luxurious to the touch. A dozen coat makers were busy hand-cutting, smearing glue with their fingers, and rolling; the same men who pieced together the old macs for sewage workers now handcrafting Mulberry's new line of twill macs with polka-dotted arms and hoods. As Macklemore rapped about popping tags, I imagined eye-popping price tags; the polka-dot Mulberry would retail the following spring for $3,000.

My tour guide was the production manager Willie Ross, a strait-laced Scotsman whom Dunko hired away from a jeans manufacturer in 1999. Since then, Ross had modernized the factory and its workplace, ventilating the glue fumes and banning the Scotch whisky and cigarettes once sucked on steadily by the coat makers. We talked about the coats' waterproof properties: Perfect, Ross told me. But the fashionistas are not looking to keep dry. "This is one hundred and twenty percent a fashion item," he said.

Merchandisers call the mac a "heritage brand," like Red Wing Shoes from America or Montblanc fountain pens from Germany. Japan has become the hottest market in the world for such brands, and the number one market globally for Mackintosh raincoats. Half of all Mackintoshes are purchased in Japan; the United Kingdom is now the second-largest market.

The dusty historical memorabilia that had collected in the Cumbernauld factory outside Glasgow when Yagi acquired the company has been shined up. Antique tailoring shears and glue pots, rollers and tapes, along with a swatch book and some of the old macs worn by U.K. police at the turn of the twentieth century, have been shipped out of Scotland for display behind shiny glass cases at the company's flagship store in Tokyo's high-end Aoyama district.

I could not help but think of the Japanese *wagasa* umbrellas—one article lost to modernity, another found. I corresponded with a Tokyo artist, Yasuko Horie, part of a new generation trying to bring *wagasa* back to the Japanese. When they see her articles, "people have the desire to touch the traditional umbrellas, and thus touch a disappearing culture," she told me. In the fickle history of the articles of rain, perhaps we'll someday tote exquisite paper umbrellas, a fashion statement to celebrate a disappearing heritage.

|||

——

AMERICAN

RAIN

SIX

FOUNDING FORECASTER

As a boy, Tom Jefferson was drawn to a peak in the foothills of the Blue Ridge Mountains that led him to imagine he could ride above the storms.

The crest rose in the Piedmont of Virginia from his father's plantation, which was known as Shadwell. Tom was born there in 1743 but spent much of his childhood elsewhere, first at another plantation downstate, and then away at school.

While he mastered Greek, Latin, and French fifty miles from home, young Jefferson looked forward to his adventures back at Shadwell, and to a land and sky he considered ideal. As the defining Blue Ridge had been formed, like him, from a crash of the European and American continents, so the region's pleasant climate was born of an auspicious collision of two global weather patterns. Cold and storms tracking across the continent from the great mountains of the west mellow at the Piedmont when they meet the warmth of the Atlantic Ocean's Gulf Stream. Temperatures turn milder, rains gentler.

On trips home to Shadwell, settled at his father's farmhouse near the Rivanna River, the long-limbed boy with freckles and sandy-red hair would set off, by foot or by horseback, about three miles to the base of his enchanting summit. It was another 867 feet to the apex, a steep climb through dense red cedars and fine-layered clouds that cleared at the top for an elysian view of the blue-gray mountains beyond.

The vista drew the boy into "the movements of nature . . . in a never-ending circle" so fully that he would obsess over the workings of plants and land, animals and atmosphere for all his life. Year upon year, Jefferson would record the date of the first whip-poor-will song, the bloom of the native dogwood trees that lit the misty mountain understory like candelabras—and, for more than five decades, almost every inch of rain that fell upon him.

It was here that, before he became one of the fathers of a nation, Jefferson conceived a more personal dominion he called Monticello, his "little mountain." "How sublime to look down into the workhouse of nature," he would write in his most intimate description of the place, "to see her clouds, hail, snow, rain, thunder, all fabricated at our feet!"

At twenty-five, Jefferson began to level the mountaintop to create his life's home. Its domed neoclassical design befitting the lofty vista and occupant, Monticello would become the most famous private residence in America. But, out of character for the draftsman of American independence who likewise led the nation's thinking on architecture and agriculture, science and invention, it was also a famously poor location for a house.

In eighteenth-century America, no one built at such heights. Even Jefferson's admiring biographer Dumas Malone wrote that his decision to settle on the little mountain "appeared to be flying in the face of common sense." The challenge for Jefferson was water. Perhaps blinded by his childhood image of riding above the storms, he grasped the problem too late.

Jefferson based his design for Monticello on the principles of the sixteenth-century Italian architect Andrea Palladio. But he ignored his Renaissance idol's first choice for locating a villa—at river's edge, to blend water's beauty with its utility for daily life.

Along the eastern edge of North America, natives had settled riverside for at least 12,000 years. More recently, ten of the thirteen original British colonies had developed along rivers—vital for transportation, quenching the garden in dry times, or powering a mill like the one Jefferson's father had built on the Rivanna near the Shadwell estate.

Slave ownership perverted the wisdoms and practicalities of the day. When it came time to sink a well at Monticello, one paid excavator and a crew of slaves dug for forty-six days, through sixty-five feet of mountain rock, to find water—more than double the depth typical for the red-clay soils of Virginia.

But there was something more at play than a planter leaving the details to labor. Jefferson was fixated on creating the perfect home and landscape to the extent that he calculated everything from the shape and number of bricks that should line his terraces to the weight of individual peas harvested from his garden. Yet so invested was he in the image of Monticello's perfection, including what he believed was the ideal human climate, he never spent as much time pondering where his water would come from. He put all his faith in the rain.

When the first British colonists set out across the Atlantic Ocean to the New World, they made the commonsense assumption that weather patterns in North America would match up by latitude to those in Europe. In fact, they were leaving one of the gentlest climates in the world for one more prone to extremes. Western Europe is milder than the eastern United States, thanks in large part to the warmth of the powerful oceanic current that also binds the two coastlines in history and culture, the Gulf Stream.

The colonists expected the land they called Virginia to soon flourish with the same crops as Spain and Portugal—olive and orange trees, sugar and spices. Instead, they sailed straight into the final and zenith years of the Little Ice Age in North America. Temperatures plunged to their lowest in the seventeenth century, freezing the Delaware River and much of Chesapeake Bay. Springtimes could bring dismal rains and dreadful floods, summers crop-killing droughts. High death rates and the failure of many of the early colonies changed European views of America and its climate like a flash flood wipes out affection for a summer shower.

By Thomas Jefferson's day, the leading scientists of Europe had come to believe the New World's climate so wet and miserable that it actually deformed the animal, plant, and human inhabitants of America—brewing puny mammals, sickly crops, and natives cursed with lack of body and facial hair and "small and feeble" reproductive organs.

This "theory of degeneracy" began with a world-famous French scientist named Georges Louis Leclerc, the Comte de Buffon, whom Charles Darwin would credit as the first to treat evolution "in a scientific spirit." Buffon's influential books, already accepted as classics by his peers, shuddered at a dank America, air heavy with vapor, vast, stagnating marshes, volumes of floodwaters "for want of proper drains or outlets," vegetation so thick its transpiration "produces immense quantities of moist and noxious exhalations."

Buffon believed America's flora and fauna were shrunken and weakened by "a niggardly sky and an unprolific land"—with the exception of reptiles and insects that thrived in the humidity. Another purported scholar of America (who, like Buffon, never stepped foot on the continent), the Dutch philosopher and geographer Cornelius de Pauw, wrote of its "putrid and death-dealing waters" over which settled "fogs of poisonous salts." America is "overrun with serpents, lizards, reptiles, and monstrous insects," grotesquely oversized, such as frogs in Louisiana that "weighed thirty-seven pounds and bellowed like calves."

Americans, and particularly Jefferson, were incensed by the atmospheric affronts. They debated the inaccuracies in college classrooms and churches, in novels and in scientific papers. They did believe, like the European scientists, that clearing the forests and cultivating the land would help moderate the New World's extremes. But Jefferson was also convinced that America's climate was already superior to Europe's. He just needed the scientific data to prove it to Buffon and to the world—detailed daily records of the rains and the winds, and the temperature and pressure in the air.

Jefferson invested in rain gauges and weathervanes, thermometers, and eventually a barometer—purchased in a Philadelphia apothecary called Sparhawk's four days after the Continental Congress adopted his Declaration of Independence. Born of patriotism and noted in his ivory-colored Weather Memorandum Book, Jefferson's meticulous weather observations became a lifelong passion that lasted from his law school days at Williamsburg to his final years at Monticello.

His experience living in France, first as U.S. trade commissioner, then as envoy to the Court of Versailles, told Jefferson that American rains were more salubrious than Europe's. Based on his observations of rainfall and cloud cover on the two continents, he hypothesized that Virginia enjoyed both more rain and more sunshine than a city such as London, where bleak skies seemed to idle in perpetual drizzle. True, it could pour buckets in Virginia. But heavy mountain rains rejuvenated. They kept the rivers and lakes filled and the groundwater robust for the well. And often, they were followed by brilliant sun and the rainbows he so admired—plunging from his vantage point at Monticello deep into the river valley below.

Jefferson's calculations were astonishingly accurate for a man who lived two hundred years before advanced weather technologies. He estimated average rainfall in his region at 47 inches a year. Modern meteorologists peg the average at 46 inches a year, falling over the course of about 116 days. In London, more rainy days—an average 133—bring less overall rain: about 33 inches, just as Jefferson expected.

But it would take a good deal more than the Weather Memorandum Book to convince Count Buffon to revise his theory of degeneracy. Jefferson's exhaustively detailed report on climate and nature in his home state, delivered to the count in 1785, was not enough. Nor was his personal testimony, when the two men met in Paris later that year and the American minister told the skeptical count that the reindeer of Europe "could walk under the belly of our moose."

Finally, Jefferson managed to pull off what may be the greatest image-repair job in history. To prove North America's animals weren't puny, he wrote to his friend General John Sullivan in New Hampshire and implored him to find and kill a bull moose in the snowbound northern woods, then ship the bone, skin, and antlers to Paris. The venture was time-consuming for Sullivan and expensive for Jefferson, but it worked. Presented with evidence of the hulking quadruped, Buffon "promised in his next volume to set these things right," according to Jefferson. But the count died soon after, and never did.

Many Jefferson biographies revel in the story of the moose. But the incredible irony in Jefferson's saga to disprove the theory of degeneracy is not found in the woods of New Hampshire. It comes, instead, from the skies over Virginia.

For many of the years he painstakingly collected rainfall data to prove the New World was not an overly stormy, damp, moist, humid, clammy, soggy, dank, showery, muggy, dewy, dripping, drizzling, sodden, oozy, murky swamp, Jefferson, his family, his crops, and most especially his slaves were suffering in drought.

The Founding Forecaster was high and dry at Monticello—wishing for rain.

For all the detail, creativity, and brilliance that went into Monticello—its carefully planned vistas, its cerebral octagonal and domed rooms, its hillside laboratories of fruits, vegetables, and other useful plants—Jefferson was overly optimistic about having the rainfall to grow the kinds of crops and live the sort of life he envisioned there.

While he was dead-on about Virginia's averages, the averages could fly out Monticello's triple-sash windows on a strong wind. The truth is that for much of Jefferson's life, lack of rain thrust Monticello into hardship. The situation would not have been so devastating had he built at the river.

Perched as it was atop the mountain, Jefferson's well, even at 65 feet, was particularly susceptible to drought. Between 1769 and 1797, he logged the workings of the well into his Weather Memorandum Book. In those years, it was dry a total of six. The architectural historian Jack McLaughlin figured out that two of the dry years fell during Jefferson's marriage, before he was widowed, "which meant that for one-fifth of her married life, Martha Jefferson had no readily available source of water at her house."

Jefferson tried to solve his water woes with rain's most enduring technology. Cisterns, ranging from the simplest clay pots to cavernous underground chambers, were the bedrock of water supply for most of human history; ancient civilizations could not have survived without them. Archaeologists have found waterproof lime plaster cisterns built into the floors of houses in villages of the Levant. During biblical times, underground cisterns not only stored water, but could serve as hiding places for fugitives or as prison cells. The Romans constructed them throughout the Empire. Terra-cotta pipes carried rain from rooftops to home cisterns for domestic use. Vaulted public cisterns collected rain from hills, clay or bronze pipes, and later the great aqueducts to feed city water supplies and the Roman baths.

Jefferson began to work on his cisterns in the first decade of the 1800s, about the same time New York City was installing them for public use. In his book *Water for Gotham*, Gerard Koeppel describes the twin 200-hogshead stone cisterns built then at the wings of city hall, fed by rainwater from the roof. By 1830, there were forty-three public cisterns throughout Manhattan. They lost favor as a source of water supply when they proved inadequate to douse the Bowery Theatre fire and other blazes that raged in the city during those times. Jefferson calculated Monticello's rooftop rain-catching potential in his Weather Memorandum Book—gallons coming in, area on the roofs, storage needed, daily supply expected. He designed four brick cubes, each eight feet square, for the four corners of Monticello's terraces. Each could hold 3,830 gallons, and bring the house an average 600 gallons of freshwater a day. He spent years struggling to seal them, finally settling on Roman cement imported from Europe. But, contrary to the evoking of Jefferson by today's rainwater-catchment industry, they never worked.

Rain has often been the bane of architects and structural engineers, who spent far more of history cursing and barricading rain than collecting it. If rain has a victory flag, it is the plastic blue tarp—flying as triumphantly over the McMansions of the U.S. Eastern Seaboard after tropical storms as atop the famously leaking flat roofs of modern architecture. As long as it took us to protect our bodies with macs, umbrellas, and other articles of rain, to this day we have not perfected the roof over our heads.

In the first modern treatise of architecture, written in Italy in the fifteenth century, architect-philosopher-poet Leon Battista Alberti cast rain as the great enemy of any building, "always prepared to wreak mischief." Rain "never fails to exploit even the least opening to do some harm," Alberti wrote. "By its subtlety it infiltrates, by softening it corrupts, and by its persistence it undermines the whole strength of the building, until it eventually brings ruin and destruction to the entire work."

Rainwater must never be allowed to leak in, Alberti implored in his *de re Aedificatoria,* or *On the Art of Building in Ten Books*—and never to stand. The key is the roof, "the first of all building elements to provide mankind with a place of shelter." A roof should always slope to throw off rain. Those large in area should be divided among several planes so rainwater can flow off in different places rather than one big gush.

The admonitions make perfect sense. But, like other obvious advice such as eating only healthy foods, the human quest for something richer makes them hard to follow. "Despite all the determination and skill that man has invested in his attempt to strengthen and reinforce [the roof] against the assaults of the weather, he has scarcely succeeded in protecting it as much as necessity demands," Alberti wrote. Six hundred years later, it's still true.

Rain can warp, swell, discolor, rust, loosen, mildew, stink, peel paint, consume wood, erode masonry, corrode metal, expand destructively when it freezes, or seep into every crack when it evaporates. The "most important organ of health" for keeping these maladies away, writes *Whole Earth Catalog* publisher Stewart Brand in his tidy book *How Buildings Learn,* is the right roof, in the right pitch and shape.

Many of the enduring roofs of history are vernacular—they speak a region's native language. The Great Coxwell Barn of Oxfordshire, England, has stood for seven centuries under a steep-pitched timber roof laid with Cotswold stone, now fuzzy with soft brown moss. Other vernacular roofs were thatched with local vegetation—reed, straw, sedge, and rushes that date to medieval Europe, or palm fans and fronds in regions near the equator. In the European countryside, the high-pitched, thatched-roof cottages of commoners were often cozier and drier than the drafty castles of nobility. Still, thatching became associated with poverty. Expanding trade routes meant architects and builders could import fancier materials such as tiles, copper, and glass. As a general rule, the showier the house and roof on top, the more likely to leak, particularly around the seams of any chimneys, skylights, balustrades, domes, and so on.

The conflict between creative design and practical rain-proofing is well captured in an old architecture-school joke: "All good architecture leaks." The bon mot traces to Frank Lloyd Wright, his devotion to organic buildings so full that rain is practically invited in. The celebrated architect's aesthetic masterpieces were notoriously plagued by leaking windows and right angles, walls, and parapets. His flat-roof designs were known to leak almost immediately. To clients who complained, Wright is said to have retorted: "That's how you can tell it's a roof." (Wright is by no means alone; Brand found some 80 percent of post-construction claims against architects were for leaks.)

Wright's exquisite Fallingwater, designed in 1935 for the Kaufmann family, straddles a waterfall in rural southwestern Pennsylvania. It lives up to its name with perennial drips, drabs, and damaged concrete. Its owner, Pittsburgh businessman Edgar Kaufmann Sr., called it a "seven-bucket building" for its leaks, and nicknamed it "Rising Mildew" for its mold.

On a "beautiful little ravine" in Pasadena, California, Wright built a small jewel, La Miniatura, in 1923 for an eternally patient widow and art collector named Alice Millard. It was the first time Wright used his new textile-block construction scheme, a way to "take that despised outcast of the building industry—the concrete block . . . find hitherto unsuspected soul in it—make it live as a thing of beauty—textured like the trees." Wright biographer Ada Louise Huxtable calls Millard "a true believer who tolerated all the trials and tribulations her house and architect provided, which were almost biblical in their nonstop intensity." As soon as the home was finished, torrential rains hit Southern California, sending floodwaters through the ravine and waterfalls through the textured blocks—inside and out of La Miniatura.

Later in the 1920s, Wright's cousin Richard Lloyd Jones Sr., publisher of the afternoon newspaper in Tulsa, Oklahoma, commissioned him to build a home there. Jones was worried about the textile blocks in an area with more rain than the West, and rightly so. Despite heroic waterproofing attempts, the home, Westhope, was perpetually damp. The roof leaked almost immediately after Jones moved in. He called in roofers to resurface, in vain. He went to his desk and placed a call: "Dammit, Frank—it's leaking on my desk!"

Wright calmly replied, "Why don't you move your desk?"

Wingspread in Racine, Wisconsin, commissioned by Herbert Fisk Johnson of Johnson Wax, had a similarly soggy fate. "For the first year or so, Johnson had workmen at the ready with putty guns whenever it threatened to rain." The story of Westhope also became the lore at Wingspread: a thunderstorm, an outraged owner, a telephone call, an accusation: "Frank . . . it is leaking right on top of my head!" And the same reply, recounts Wright biographer Meryle Secrest, "given with his usual insouciance."

Lakeland, Florida, is home to the largest collection of Frank Lloyd Wright buildings in a single site, Florida Southern College, whose president asked Wright to build a "college of tomorrow" to boost enrollment during the Depression. The dozen Wright structures are known collectively as "Child of the Sun"—"out of the ground and into the light, a child of the sun," as Wright described his work there. But Lakeland is known as much for its rain; it is the thunderstorm capital of the United States, with storms rolling in about a hundred afternoons each year. When I toured the Wright buildings during a spectacular sunset in 2013, the Annie Pfeiffer Chapel smelled of mildew. Its triumphant glass panels were clouded with trapped moisture. The building's textile blocks had leaked for decades. The college has landed a considerable grant to replace them; the blocks crafted on site by students who needed work in the late 1930s will be replaced by high-pressure mold blocks, professionally manufactured.

Mark Tlachac, director of the Frank Lloyd Wright Visitor's Center at Florida Southern, told me it would be a mistake to think that Wright overlooked the rain, or had any arrogance toward climate. To the contrary, Wright loved rain—really, any showy weather—and wanted those who inhabited his buildings to feel its drama. The Florida Southern campus is criss-crossed with covered walkways hanging several inches on either side of the sidewalks so that crowds can walk in the rain and enjoy it without getting wet. Wright often eschewed rain gutters because he wanted to celebrate rain rather than hide it. "He loved the look of icicles hanging from a roof in the Northeast, and the look of rain running off in the South," Tlachac told me. "He loved nature, and he wanted to live with nature. Sealing it off wasn't so much of a priority."

Sealing was the priority and ultimately the insurmountable problem for rain-loving Thomas Jefferson and his cisterns. The Roman cement was supposed to be waterproof, but the cisterns never would seal completely. Only two of the four held water reliably as years of scant rainfall stretched on.

Jefferson's rich correspondence makes little mention of the practical difficulties of drought for his family, or for his slaves, who, faced with a dry well, would have been forced to haul water for the house and its terraced garden up a steep incline from the nearest spring, halfway down the mountain. He fretted most in his letters about the lack of rain for quenching his fields—and those crops beginning to spread across the young nation:

> *The drought is excessive. From the middle of October to the middle of December, not enough rain to lay the dust. A few days ago there fell a small rain, but the succeeding cold has probably prevented it from sprouting the grain sown during the drought.*
>
> —JEFFERSON *to* JAMES MADISON, *December 1796*

> *The effects of drought are beyond anything known here since 1755. There will not be 10,000 hogsheads of tobacco made in the State. If it should rain plentifully within a week, the corn in rich lands may form nubbings; all the old field corn is past recovery, and will not yield a single ear. This constitutes the bulk of our crop; there will be no fodder. The potatoes are generally dead.*
>
> —JEFFERSON *to* ALBERT GALLATIN, *August 1806*

> *We are suffering here, both in the gathered and the growing crop. The lowness of the river, and great quantity of produce brought to Milton this year, render it almost impossible to get our crops to market. . . . Everything is in distress for the want of rain.*
>
> —JEFFERSON *to* JAMES MONROE, *May 1811*

> *We are here laboring under the most extreme drought ever remembered at this season. We have had but one rain to lay the dust in two months. That was a good one, but was three weeks ago. Corn is but a few inches high & dying. Oats will not yield their seed. Of wheat the hard winter & fly leave us about ⅔ of an ordinary crop so that, in the lotteries of human life, you see that even farming is but a gamble.*
>
> —JEFFERSON *to* JAMES MONROE, *June 1813*

Perfectly elegant yet usually dry, Jefferson's cisterns are unfilled symbols of his ambition for the spread of family farms westward and throughout the young nation's vast interior. Jefferson had an almost religious vision for the cultivation of America's lands beyond the Blue Ridge, and considered the advance of independent farms the most important task facing the first generation of U.S. citizens. Self-reliant farmers would tame the wilderness; build the nation's economy and democratic principles; and make Americans a "tranquil, healthy and independent" people. He believed it despite early reports of desert conditions coming in from the West.

Jefferson never regretted building on the little mountain that enchanted his boyhood. Despite all the hardships, history deemed the selection exceedingly original, and impressive to every visitor during his lifetime and centuries later. In 1816, one of those visitors, U.S. Attorney General Richard Rush, made the pilgrimage to what he called the mecca of Monticello to visit now-aging Jefferson, an old friend of his father. He would marvel that the fog never rose to the top of the mountain, where Jefferson, "in genius, in elevation, in the habits and enjoyments of his life . . . is wonderfully lifted above most mortals."

"If it had not been called Monticello," Rush wrote, "I would call it Olympus, and Jove its occupant."

It was a telling characterization. Mount Olympus had its gate of clouds. And Jove is Jupiter—king of the gods in ancient Roman mythology who was also god of the sky, rain, thunder, and lightning. In his role as rainmaker: Jupiter Pluvius.

Pluvius personified—*With what majesty do we there ride above the storms! How sublime to look down into the workhouse of nature, to see her clouds, hail, snow, rain, thunder, all fabricated at our feet!*—Jefferson's relationship with rain is also America's. His careful weather logs reflect the quest of the scientific mind—the mind that invented practical gadgets and developed a philosophy so rational that "the American experiment would prove that men can be governed by reason and reason alone." Yet he held a certain defiance for the reality of climate. With the same audacity that draws us to develop in the wettest areas or farm the driest, to put our homes in floodplains or in the path of hurricanes,

Jefferson chose to build a home with poetic views rather than potable water. He was devoted to his vision that small yeoman farmers would build the country's wild interior and its economy—before he knew whether they would have enough rainfall to farm. In too many years, they would not.

At the turn of the nineteenth century, most Americans still lived within fifty miles of the Atlantic Ocean. In his 1801 inaugural address after he was elected third president of the United States, Jefferson expressed his vision for "a rising nation, spread over a wide and fruitful land . . . advancing rapidly to destinies beyond the reach of mortal eye." Two years later, he sold Congress on his plan to fund an expedition all the way to the "Western Ocean."

When Jefferson sent Captain Meriwether Lewis and William Clark to explore the economic potential of the West and its life, waters, and lands, he made a special climatic appeal. His official instructions emphasized that they should observe and record "climate as characterized by the thermometer, by the proportion of rainy, cloudy and clear days, by lightning, hail, snow, ice, by the access & recess of frost, by the winds prevailing at different seasons, the dates at which particular plants put forth or lose their flowers, or leaf, times of appearance of particular birds, reptiles or insects."

In summer 1804, Lewis and Clark topped a Nebraska hill to their first wide-angle view of America's great, grassy middle. Clark noted a "bound less" and "parched" prairie. Seventy-five years later, even Walt Whitman could not adequately describe the treeless, windswept prairies between the Midwest and the Rocky Mountains: "One wants new words in writing about these plains, and all the inland American West—the terms, *far, large, vast, &c.,* are insufficient."

From the Mississippi River west to the Rockies and from Canada's Saskatchewan River to what is now northern Texas, endless prairie grasses bent in ceaseless winds. On the eastern side, which might see twenty inches of rainfall in a good year, lush grasses could grow higher than a wagon. In the West, tougher, shorter grasses muscling their way up through gravelly sands saw ten inches at best. Either way, the region seemed no place to settle. Native Americans lived well on the plains; they did not depend on capricious rainfall, but instead followed the bison.

In July 1804, Clark had complained of "the wind Blustering and hard from the South all day which blowed the clouds of Sand in Such a manner that I could not complete my p[l]an in the tent." The following month, too, "the wind blew hard and raised Sands . . . in such Clouds that we could scarcely [see]." A heat wave seared so harshly that Captain Lewis's eager black dog, a Newfoundland named Seaman, became too fatigued to walk.

Just a few years behind Lewis and Clark, in 1806 and 1807, a young military officer with an aristocratic face named Zebulon Montgomery Pike Jr. led an expedition to explore the southern portion of the Louisiana Purchase and find the headwaters of the Red River. His report referred not to a Great Plain, but a Great American Desert. Settlement was out of the question. The problem wasn't Indians, but the cottonmouth dry:

> But here a barren soil, parched and dried up for eight months in the year, presents neither moisture nor nutrition sufficient, to nourish the timber. These vast plains of the western hemisphere, may become in time equally celebrated as the sandy deserts of Africa; for I saw in my route in various places, tracts of many leagues, where the wind had thrown up the sand, in all the fanciful forms of the ocean's rolling waves, and on which not a speck of vegetable matter existed . . . But from these immense prairies may arise one great advantage to the United States . . . The restriction of our population to some certain limits.

If Pike's wasn't a bleak enough admonition, another army officer, Stephen Harriman Long, offered an even drearier one ten years later. Long led the first scientific exploration up Nebraska's Platte River in

1821, a trip so harsh that he and his men had to kill, roast, and eat their own horses to survive it. He described the Great Plains from Nebraska to Oklahoma as so parched they were "unfit for cultivation and of course uninhabitable by people depending on agriculture.

"The traveller who shall at any time have traversed its desolate sands will, we think, join us in the wish that this region may forever remain the unmolested haunt of the native hunter, the bison and the jackal."

Pike's and Long's antiexpansionist wishes were not to be. For in the late 1860s, rain began to pound the plains, beading up and running off the hard crust. As the ground soaked in new life, the once-desolate sands turned colors—from drab dust to gorgeous green. Now a long humid cycle brought above-average rainfall to the Great American Desert. The name, in fact, disappeared from maps by the 1880s.

The showers kept coming, year after year. And so did the settlers, many of them families. They came to be called sodbusters for their prairie-breaking work. Atop the promising climate, they were lured by cheap land made available by the federal government, now hard-selling Jefferson's vision of a nation of independent yeoman farmers.

For years, southern states had successfully fended off western settlement incentives in Congress. But secession assured their passage. Western boosters pushed the first Homestead Act through in 1862. Anyone willing to head west, build a home, and grow crops could have 160 acres all his or her own. (Single women and widows were eligible for the land, priced at a dollar and twenty-five cents an acre.)

While 160 acres may have been sufficient for an eastern farmer, it was not as good as it sounded for small farmers on the dry plains, where scarce natural vegetation also made raising livestock difficult. In 1878, Major John Wesley Powell, the adventuresome, one-armed Civil War hero who headed the U.S. Geological Survey, warned Congress in his *Report on the Lands of the Arid Region* that scant rainfall made the needs of small farmers in the West entirely different from those in the East. A line down the middle of the Dakotas, Nebraska, Kansas, Oklahoma, and Texas, the 100th meridian, divided lands with enough rainfall for small farms—twenty inches was the conventional wisdom—from those where farmers should instead organize communally, around watersheds.

Truth was, much of the arid federal land being doled out to hopeful settlers would simply not support small farms. Even in what Powell called the "Sub-humid Region," along the 100th meridian itself, farmers might thrive some years, but then face "disastrous droughts."

Congressmen didn't buy it. Powell's warnings were not welcome in a nation bent on expansion. The environmental historian Donald Worster, eminent chronicler of the water empire of the American West, said western congressmen in particular saw Powell's ideas as radical, with "too much planning, too much regulation, too much community control. This is not the American way."

Through the 1870s and early '80s, thousands of homesteaders rushed to Dakota, Nebraska, Kansas, and beyond. They replaced ashen clumps of sagebrush with leafy cornfields. They earned record crop prices. They sent elatedly for relatives back east. They did not know these rain-filled years were outliers: sheer meteorological coincidence.

Historians have always pointed to land as the basis for Thomas Jefferson's dream of a nation, realized through yeoman farmers one patch of soil at a time. But it was not land that left his vision unfulfilled. To the west, lack of rain would be the spoiler.

SEVEN

RAIN FOLLOWS THE PLOW

n his love letters to Mattie, and later, in those signed "Your Affectionate Husband," Uriah Oblinger would often write of rain.

A Union veteran with a wiry beard and high cheekbones, Uriah first journeyed from his home in Onward, Indiana, to the West in 1866, "in quest of a peaceful and permanent home" for himself and Mattie. Hard rains soaked his campsites and slowed his progress. His horses strained and his wagon wheels caught in the thick mud glutting the trail. Not long after he'd crossed from Iowa into Missouri on the Turkey River ferry, one protesting wheel popped a kingbolt, forcing him to retreat for repairs and lose a day in a forsaken-looking town.

Great deluges scared the colts, chilled his bones, and made his piles all the more miserable. Yet, wretched as it could be, Uriah was convinced that rain was a gift from God, sent to endow the soils and fill the young nation with small farms. This was as clear as the rainbow that had greeted him at the Mississippi River and its great limestone gates of western promise:

"Sun rose beautiful amid the grand old Mississippi bluffs," he wrote to Mattie on the night before he crossed the Mississippi River. "A beautiful Rainbow at Sunrise sprinkling some with Sun shining bright . . ."

Uriah's initial ramblings failed to produce a peaceful and permanent home. He and Mattie married back in Indiana in 1869 despite her father's reservations about her tumbleweed of a correspondent. After three harvests on a rented farm, Uriah journeyed west once more, this time with his brothers-in-law, to stake homesteads in Nebraska.

Again, rain soaked him. In Illinois, he and Mattie's brothers steered their wagon team through downpours, camped in storms, and at one point passed fountainheads bubbling so vigorously "all we had to do to get water was hold our bucket und[er] and let it fill without lab[o]r."

One September morning, Uriah sat upon his provision box in a most welcome ray of sun, and wrote of all-night rainfall in a letter to Mattie and his newborn girl, Ella.

"Is it raining at your house last night and this morning?"

He wanted not one cloud of sorrow to dim their happiness.

During their journey, Uriah and his brothers-in-law met many fellow emigrants in wagons, often at river crossings awaiting a ferry. By the time they reached Nebraska in October 1872, it seemed that every other young man in the East had the same notion to settle in the five-year-old state. They had to forge well beyond the city of Lincoln to find what they hoped would be unclaimed acreage. About seventy miles past, down south toward the Kansas border, Uriah chose what he thought was the perfect 160-acre claim. The prairie land had a creek at the corner, with plenty of water for stock. Half the land had already been homesteaded, by an unfortunate young man who'd broken his back and could no longer tend it. But by the time Uriah was able to return to the Land Office in Lincoln, someone else had already staked the property.

Uriah was down to his last few dollars. He felt hungry all the time, and so seemed his old horse, Nellie. Her ribs poked out, he wrote to Mattie. He traded his wagon for one not so nice, and his shotgun as well, for the cash he needed to set out again. Searching the treeless, flat prairie for a stake, he convinced himself that a creek wasn't so important; in fact, it would get in the way of the plow. He and Mattie would dig a deep well and haul water up in buckets with a crank-and-rope windlass.

Fillmore County, Nebraska, after all, lay in the heart of what would come to be called the Rainwater Basin. In times of rain, 4,000 playa lakes, or shallow wetlands, would fill out across 100,000 acres of grassland, then percolate exceedingly slowly into an ancient aquifer, now known as the Ogallala, which sat undiscovered until the invention of the diesel-powered water pump. (When the conquistador Francisco Vázquez de Coronado pushed his men into the High Plains from New Spain in the sixteenth century to search for the Seven Cities of Gold, they went near mad with thirst—never knowing of that vast freshwater sea beneath their feet.)

To be sure, Fillmore County farmers still prayed for rain. Dust continued to blow across Nebraska, fierce as the battles Uriah had seen in the cavalry. But this was no desert. Recurrent great storms were proof of that. Just as they were proof that settling this once-desolate land was all part of God's plan.

In New York, the artist John Gast had just finished his allegorical painting *American Progress*. An enormous, creamy-skinned beauty in a diaphanous gown floats westward across the prairie. She is the goddess America, schoolbook in her right hand, a stream of telegraph wires in her left, leading the nation to its manifest destiny of a fully civilized continent. Nineteenth-century settlers held an abiding belief in Manifest Destiny; that they had been specially chosen by God to advance America and the agrarian dream. Gast's beauty brings light and progress from the East. Ahead of her, storm clouds darken the western horizon, where frightened natives and bison flee. Behind her, the mighty Mississippi carries busy ships, and trains chug forward— every means of transportation possible to move the new nation west.

Gast's painting would be widely distributed in promotional literature urging frontier settlement, and in prints that hung on the walls of homesteaders. Uriah was not likely to have seen it by that spring when, his claim secured, he wrote to Mattie that she should begin preparations for her journey to Nebraska.

The country was settling fast. One hundred and sixty acres belonged to them, and soon, Uriah would harvest corn to sell for train fare. "Surely the hand of Providence must be in this," said Uriah. "It seems this desert as it has been termed so long has been specially reserved for the poor of our land to find a place to dwell in and where they can find a home."

As settlers began to push farther and farther west across Nebraska and Kansas, a curious phenomenon occurred. The rain seemed to follow like an obedient hound.

When settlement moved beyond the 98th meridian, so did the rain, as if the two were linked. Could it be? A cause-and-effect notion swirled first as superstition. Then, it caught hold among the homesteaders. Surely this was more proof of His providence, and their destiny to settle the continent through to the Pacific Ocean.

Railroad bosses loved the theory. They launched propaganda campaigns that claimed settlement was turning the Great American Desert into a "rain belt." Town boosters, the federal government, and even some scientists—though many men of science tried to warn that it was hokum—jumped on the bandwagon. At the just-established University of Nebraska, one pair of scientists legitimized the theory with an explanation. Samuel Aughey was a natural sciences professor and the state's geologist; Charles Dana Wilber served in the university's Department of Geology and Mineralogy. Both men also worked as advisers to the railroads, which could not stretch their tracks from the eastern cities to the West Coast without boomtowns in between.

Aughey reported on several "facts of nature" to prove rain was increasing on the plains: much taller grasses than those observed by Lewis and Clark in 1804; newly bubbling spring-fed creeks bone dry since '64; and most important, ten years' worth of greater rainfall than that recorded in the decade prior.

If the idea of a long-term climate cycle occurred to Aughey, he did not share it.

His reason for the rain, instead, was "the great increase in the absorptive power of the soil, wrought by cultivation." Wilber explained that Nebraska's landscape had been "pelted by the elements and trodden by millions of buffalo and other wild animals, until the naturally rich soil became as compact as a floor." These soils could not hold rain; any drops that fell would roll away like a child's marble lost in dust under the bed. Once the dutiful settlers broke up land with their plows, however, rainfall could seep down, then return to the atmosphere overhead. The more soil cultivated, the more moisture captured. More moisture, more evaporation. More evaporation, more rain.

"Rain follows the plow."

The theory stuck like wet grass. The railroad boosters picked it up as a slogan, using it in promotional literature to hawk settlement in the Great Plains region well into the twentieth century.

Uriah took up his plow to build a house for Mattie and Baby Ella. He dug into the tough prairie grasses four inches deep, and cut strips a foot wide and up to four feet long. He hauled the heavy slabs to his home site—a gentle slope with a beautiful view and morning sun. He laid them, grass side down, into rows that would become his walls. Three rows of sod, tucked side-by-side, would create a barrier thick enough to withstand Nebraska's constantly blowing winds. As he built up his walls, staggering the sod strips like bricks, he fretted about the recent lack of rain, and those winds.

He hoped Mattie could stand them. Moreover, the dust they sent flying. These were not the gentle evening zephyrs of Indiana. The gusts were so strong that Uriah couldn't keep his hat on as he hauled more strips of earth. One devilish gust nearly blew him off a wall where he was perched to place his sod.

As for the dry conditions, wheat sowed a month before was barely visible. Uriah worried about his early rose potatoes. Fillmore County needed a storm, and soon.

When he finished the walls after nine days' work, his and Mattie's new home was about the size of a dining room in the three-story houses that had become all the rage in the East. On Easter Sunday, he wrote Mattie and Ella how he missed them, and how they'd be separated only two or three Sundays more, though adding a caution. "Ma, there will be many privations to endure." Atop the drought and the wind, he had begun to wish he and Mattie would not have to dig quite so far down to reach water. But, as long as they were blessed with health and strength, "we will succeed by & by then we can live happily together."

That evening, as if to answer Uriah's prayers, a whip-crack of thunder sounded from the west. At first, the rain fell steadily. Overnight, it turned violent, stirred up by gale-force winds that knocked out telegraph service to most of Nebraska. In the morning, the rain turned to sleet. The sleet turned to snow. The gales blew it howling white.

The settlers, thinking winter was long over, had been waiting until summer to fix or finish stables. Now they could hardly stand as they ventured out to check on neighbors and livestock. The white swirls blinded their eyes. They had to shout loudly to be heard. The storm raged for eighty hours. In a wide swath across Nebraska, thousands of horses, pigs, oxen, cows, and their spring newborns froze to death. Several ranchers died, as did some hundred bison-hide hunters surprised by the storm. And so did a handful of settlers throughout Nebraska's counties, some heartbreakingly close to home but unable to see their way out of unfamiliar drifts. Never known was the death toll among the ever-larger groups of emigrants headed in on the trails, sheltered only by wagons or tents.

The storm was among a succession of capricious weather events, including the lasting above-average rainfall, on the plains in the 1870s and '80s. The time and the place birthed a new word: *blizzard*. It was thought to come from the German *blitz* —"lightning"—or from a similar root, the English dialect prefix *bliz*—"violent in action."

News of what came to be known as the Easter Sunday Blizzard took a while to trickle out. At the *Omaha Republican*, newspaper editors, no doubt pressed by the railroad men, "advised against telling to the world the whole truth about the storm of 1873." The young state was starting to overcome its maligned reputation as a desert. Emigrants were streaming in. The railroads soon would be able to complete tracks clear to California. On the plains, cities would boom. In the fields, corn and wheat would flourish. Homesteaders could finally win the rewards of their risk. The last thing Nebraska needed was bad press.

Uriah Oblinger engaged in some bowdlerization of his own. He described to Mattie the beauty of the white banks left by the storm, and assured her that, while their home was filled with snow, "it was not near as bad as I expected."

Mattie's brother, Sam, did not share Uriah's optimism. He would move back home to Indiana before year's end. Sam "seems some little discouraged since witnessing the storm," Uriah wrote, "but there is no use of that for it is the most terrible storm ever witnessed here and may never occur again."

Mattie and Ella arrived at their Nebraska home in May 1873 in "considerable of rain." It was too wet to plant corn, Mattie wrote her family in her first letter home to Indiana. The sod house was not quite as convenient as a nice frame. And Mattie wished for floors. But she accepted it all with great humor. "I tell you in solid earnest I never enjoyed my self better," she wrote in June. It wasn't the novelty, but the feeling of being on her own with her husband: "every lick we strike is for our selves."

Uriah had planted twenty-three acres of corn. Sprouting in the garden were tomatoes and cucumbers, beans and squashes, 130 cabbages, melons, beets, and the nicest early rose potatoes in the neighborhood. Mattie gushed to her parents about the rich green color of the prairie, the variety of wildflowers. "We have had an immense lot of rain here this season," she wrote, "but I guess this has been the general complaint every where."

Uriah and Mattie and the other settlers did not forget they were living on what was once called a Great Desert, uninhabitable until now. They truly believed they were changing the climate. "It seems as though we are destined to help make what was once called the Great American Desert blossom as the rose," Uriah declared.

Within five years, Uriah's farm blossomed, too, with acres of plump wheat and corn, oats and barley; a mess of hogs; cows to provide milk for what were now three daughters. For the first time in his life, Uriah was financially comfortable. Inside the sod house, Mattie made plans to plaster the walls. Outside, in the expanding orchard, the couple planted peaches, walnut, cottonwood, willow, and plum trees. The neighborhood built a school, where Ella, their first child, became champion speller.

And through it all, rain.

Along with the conviction that "rain follows the plow," another common belief held that the mechanical trappings of settlement could shake down rain. Some Americans were convinced that rail and telegraph lines expanding to the West were triggering violent thunderstorms. The financier Jay Gould, at a time when he controlled both the Union Pacific and Missouri Pacific railroad companies and Western Union Telegraph, made the claim that railroad and telegraph construction was expanding the nation's rainy district by some twenty miles west a year.

In reality, dry was normal. But Mattie would never know it in the seven years she worked to build a life on the plains with Uriah and their three girls. She died in childbirth in February 1880, along with their infant son. Uriah was left with nine-year-old Ella, Stella, five, and Maggie, two.

That spring was dry: day-and-night different from the year Mattie had arrived on the plains. Uriah's crops failed. He blamed himself as much as the lack of rain. "I hardly know how to manage," he wrote to Mattie's parents in the fall, asking that they take one of the children. "I feel that it is not possible or right for me to go through another season as I have this one, for I cannot do justice to myself or family this way."

Soon the wet years ended altogether. The customary arid climate returned. Had Uriah and his fellow settlers been able to read the rings of western Nebraska's red cedars and ponderosa pines, they would have seen that rainfall years with twenty to thirty inches were few and far between. The annual average was closer to thirteen. Stretches with no rain were part of the cycle, too.

The late 1880s and '90s were a time of extraordinary hardship for the settlers who'd come to farm under cool rains. The plains burned to brown once more. The shining corn leaves curled and scorched and died in the fields. Hot winds swirled. Cinch bugs swarmed. Banks foreclosed on homes and businesses. Between 1888 and 1892, half the population of western Kansas and Nebraska gave up and moved back east.

As the settlers who remained faced first hunger, then death, it was no longer easy to accept weather as providence. At the *Lincoln State Journal,* a young reporter named Stephen Crane decried the mercilessness: "The farmers helpless, with no weapon against this terrible and inscrutable wrath of nature, were spectators at the strangling of their hopes, their ambitions, all that they could look to from their labor. It was as if upon the massive altar of the earth, their homes and their families were being offered in sacrifice to the wrath of some blind and pitiless deity."

Of course, the inscrutable wrath derived not from nature but from the growing national conceit that Americans were destined to bridle nature and even the rain: bringing rain to those parts of the country deemed too dry, holding it back from those too wet. This would be the paradox of American settlement, not only on the Plains, but from the arid California coastline to the vanishing swamps of Florida to the Mississippi River. Along the Mississippi, in the third-largest river basin in the world, the same federal government that doled out 160-acre homestead plots to small farmers in a land too dry for corn built levees promised to withstand floodwaters in a land too wet for cities or cotton.

In a generation, one promise would blow away in dust-filled clouds. The other would break through the levees of the Mississippi River.

The Mississippi flows on a long, familiar journey from Lake Itasca, Minnesota, for 2,350 miles to the Gulf of Mexico. But it is really much larger than the curving line on a U.S. map or a dark and silty steamboat passage through a Mark Twain book. The Mississippi River watershed is America's great rainwater catchment. Wide at the top like a funnel, the watershed spreads from Alberta and Saskatchewan, Canada, over all or parts of thirty-one states before it gradually narrows to its spout at the Gulf of Mexico. There, the river finishes its work of spreading rain through the prairies and forests, ultimately delivering rain home to the sea.

The rain overflow of more than a million square miles of the continent, along with melt from the snow that blankets the northern plains for much of winter, pours into the catchment year after year. For millennia, the Mississippi has caught all the runoff from the broad middle of America—draining 42 percent of the United States in spring floods that inundate endless floodplains, bottomland hardwood forests, and tea-colored swamps.

Flooding is what makes the Mississippi River what it is. This is a mantra of my friend Christine Klein, a water law professor and Mississippi River author who grew up in the heart of the watershed, along the floodplain of the Missouri River just above its confluence with the Mississippi. "Without floods, the river would be just a ditch, and its floodplains arid, lifeless land."

In the sixteenth century, Spanish conquistadors led by Hernando de Soto, the first Europeans to travel across the Mississippi and its delta, described the landscape during spring flooding as an "inland sea." Native people "built their houses on the high land, and where there is none, they raise mounds by hand, especially for the houses of the chiefs; with galleries around the four sides of the house where they store their food and other supplies, and here they take refuge from the great floods."

When French settlers floated down the Mississippi into the territory they claimed and named for King Louis XIV, Louisiana, they found the fertile soils irresistible for farming, as was the river's potential for commerce and transportation. After President Jefferson orchestrated the Louisiana Purchase in 1803, giving the United States the Great Plains along with most of the Mississippi's rainwater catchment, we spent the next century building cities in the floodplains and turning the bottomlands to cotton lands. Annual flood disasters were as predictable as drought on the plains.

At first, local leaders and farmers tried to build their own levees to deal with the inundations. The structures were no match for the floodwaters of a million square miles. So many floods drowned people and their places, it is a wonder anyone rebuilt. In 1903, floodwaters jumped the banks of the Kansas and Missouri rivers, both tributaries of the Mississippi. The waters rushed to find their floodplain, now covered by Kansas City's central business district and neighborhoods. Twenty-two thousand people were flooded out of their homes. Sixteen of seventeen bridges spanning the lower Kansas River washed downstream. Over in Topeka, the state was just finishing its capitol building. Twelve feet of water swamped the capital city and twenty-four Topekans drowned.

In the nineteenth century, Americans still turned to God to save them from floodwaters; hung a snakeskin on the fence to break a drought; or, as we'll learn, forked over $500 to a man called the Rain Wizard to tap his secret formula for storm. Now, hardened by tragedies like the Flood of 1903 and the Great Flood of 1913 that drowned parts of twenty states, Americans were less likely to look upon such disasters as God's providence. As floods regularly soaked capital-infused cities from Cairo, Illinois, to New Orleans, and wealthy cotton plantations throughout the lower watershed, faith and fear turned to determination—and audacity.

The environmental historian Donald Worster has described flood-control efforts as having a "limited, ambiguous impact on the structure of society and power," whereas irrigation's impact is constant and pervasive. But mass drainage and flood control fated mid-America and the wet East as profoundly as irrigation, dams, and diversions shaped the arid West—especially in the tremendous watershed that covered nearly half the nation. Sending rain to those arid regions in the West that didn't get enough, and barricading rain from its natural catchments in the East, each side of the 100th meridian built its version of what the German-American historian Karl Wittfogel called a "hydraulic society," faith and fate now placed in centralized government power over water.

Responding to a decade of deluge, Congress in 1917 brought the federal government into the work of flood control, agreeing to fund and bolster levees up and down the Mississippi. Existing levees would be strengthened and raised three feet above the high-water mark; new ones would be built to federal standards. Ten years later, so many local, state, and federal barricades fortified the river that it was already America's "artificial gut"—William Faulkner's words.

"Now in condition to prevent the destructive effects of floods," the chief of the U.S. Army Corps of Engineers boasted. He uttered the words one year before the dark and silty waters of the Mississippi unleashed America's worst environmental disaster up to that time.

In the 1910s and '20s, another cycle of uncommon rains returned to the plains. In fall 1926, heavy storms bedeviled much of the country. August marked the first television transmission of a weather map, beamed from the navy's high-power radio station in Arlington, Virginia, to the Weather Bureau Office in Washington, although electronic storm and flood warnings for the masses were still a long way off.

Later that month, a monstrous low-pressure system dumped rain on Nebraska, South Dakota, Kansas, and Oklahoma. Rain on the plains can fall in a fury, like one of the era's most popular ads: "When it rains, it pours."*

The huge rainstorm moved east into Iowa and Missouri, then to Illinois, Indiana, Kentucky, and Ohio, lingering for days. Another moisture-heavy system followed, dropping more rain. Yet another was close behind. The storms sent dozens of streams and rivers over their banks and began to wash out bridges and railroads. By Christmas, flooding had left thousands homeless. Gauge readings on the Ohio, Missouri, and Mississippi rivers were the highest ever known.

As the calendar turned to 1927, storms continued to rumble over almost the entire Mississippi watershed, so incessant that rain made newspaper front pages around the country. Down in New Orleans, deluges washed out some of the biggest Mardi Gras parades of the season. "Proteus, Monarch of the Sea, with his parade less than half completed, decided the downpour was too heavy and turned his pageant back to the den," the *Times-Picayune* reported.

In the spring, five epic storms swept across the Mississippi watershed, each greater than any seen in the preceding ten years. The largest hit on Good Friday in April. It soaked a hundred thousand square miles from Cairo, Illinois, to the Gulf of Mexico. New Orleans broke records with 15 inches of rain in eighteen hours. (The slow-moving gullywasher Hurricane Isaac dropped 8 inches of rain on the Crescent City in a day in 2012; Hurricane Katrina, 4½ inches on the same day in 2005.)

Through late April and early May, the Mississippi's floodwaters rose phenomenally. In two separate waves, flood crests in the newly fortified river topped all previous records—approaching sixty feet above sea level. As John M. Barry explained in his chilling history of

---

* The saying was an ad rather than an adage, developed for Morton Salt. Prior to 1911, table salt had the notorious problem of caking in rainy weather. That year, Morton began adding the anticaking agent magnesium carbonate to its salt. Executives wanted to emphasize the idea that it would pour even in the damp. The original pitch, "Even in rainy weather, it flows freely," proved clunky. "When it rains, it pours" was catchy—and made all the more so by the umbrella girl still emblazoned on Morton's blue cylinder.

the 1927 flood, *Rising Tide,* the levees, built as high as forty feet, created a man-made catastrophe far worse than any natural flood could
have wrought. "These heights changed the equations of force along the
river," Barry wrote. "Without levees, even a great flood—a great 'high
water'—meant only a gradual and gentle rising and spreading of water.
But if a levee towering as high as a four-story building gave way, the
river could explode upon the land with the power and suddenness of a
dam bursting."

Levee by levee, the illusion of safety behind the government barricades began to crack. On April 15, the first length of levee, 1,200 feet
long, collapsed just south of Cairo. Across the Delta, African American plantation workers and sharecroppers were forced to the levees to
fill sandbags. Thousands of men worked desperately to save the levee at
the Mounds, Mississippi, ferry landing. Held at gunpoint, black laborers had to keep filling sandbags when everyone could hear the warning
roar of the water in their ears and feel the barricade shaking under
their feet. No one knows how many were swept to their deaths when
the Mounds levee broke. The *Jackson Clarion-Ledger* reported, "Refugees coming into Jackson last night from Greenville declare there is
not the slightest doubt in their minds that several hundred negro plantation workers lost their lives in the great sweep of water."

The muddy torrents crashed into the Delta with more than double
the force of flood-stage Niagara Falls, and inundated more than 2.3
million acres. It was more water than the entire upper Mississippi had
ever carried, more than it has ever carried since. People scrambled onto
the roofs of houses, then the houses washed away. They took refuge
in the tops of trees, then the trees gave way. To take pressure off the
levees protecting New Orleans, authorities dynamited a levee downstream in Caernarvon, Louisiana, which flooded most of St. Bernard
and Plaquemines parishes. The homes and fields of the poor living to
the east and south of New Orleans were sacrificed for what the city
fathers considered a greater good. (The dynamiting left scars so deep
that many living in the Lower Ninth Ward when Hurricane Katrina

barreled into New Orleans in 2005 insisted the levees had been dynamited once more to save the wealthier, whiter sections of the city.)

At its widest point, north of Vicksburg, the Mississippi again
turned into an inland sea—one hundred miles across. Some 637,000
people, many in the Delta, were left homeless. The river and its tributaries took lives from Virginia to Oklahoma. The death toll remains
unknown. Barry wrote that the government officially reported 500
dead, but a disaster expert who visited the flooded area estimated more
than 1,000 perished in the state of Mississippi alone. The last of the
floodwaters did not recede until autumn, leaving behind a mud-caked
and barren landscape.

The Great Mississippi Flood of 1927 was "not a natural disaster,"
said Governor Gifford Pinchot of Pennsylvania as he toured the destruction. As President Theodore Roosevelt's chief conservation officer,
Pinchot, a forester, had unsuccessfully tried to convince Congress and
the U.S. Army Corps of Engineers that flood control should work in
hand with nature. He, Roosevelt, and other Progressives had fought
the "levees only" strategy of the Corps in favor of a varied approach
that included preserving natural flood areas, building reservoirs, and
reforesting vast tracts of the upper watershed.

Congress ignored the advice—just as it had ignored John Wesley
Powell's on the arid lands the century before. In fact, Powell's *Arid
Lands* report of 1878 had mentioned the problem of floods worsening
in the East, including the Mississippi Valley, and warned: "The removal of forests and of prairie grasses is believed to facilitate the rapid
discharge from the land of water from rain and melted snow, and to
diminish the amount stored in the soil."

The hydraulic society had come to believe human-made systems
could best those shaped by climate, water, and land. In America's great
rainwater catchment, we settled in the floodplain, removed the great
soaking grasses from the prairies, razed enormous tracts of flood-
absorbing forest in the watershed, and built levees to terrifying heights.

"It's a man-made disaster," Gifford Pinchot concluded. The same would be said only three years later, when Americans again endured the worst environmental catastrophe up to that time.

On the plains, the outlier storms and floods of the 1910s and '20s seemed to wash away memory of the drought that had seared Uriah's generation. The heavy rains, along with rising wheat prices, war in Europe, and generous federal farm policies, created a second land boom on the plains. This time, many more hopeful farmers staked out acreage, drawn by the still-simmering Jeffersonian dream of land-based democracy and capitalism.

The second generation of sodbusters drove in on gasoline-powered tractors, pulling an array of disk plows. In 1925, Ford Motor Company's tractor division, Fordson, rolled its 500,000th tractor off the Rouge River assembly line in Detroit. Fordsons, along with John Deere's upstart Model Ds, helped make these farmers much more efficient than their predecessors at stripping the tough native grasses from the land.

The farmers—both homesteaders migrating in from the East and "suitcase farmers" moving in for quick profit—turned more than five million acres of grassland into golden wheat crops.

When the Depression hit and wheat prices crashed, they tore up still more of the protective grasses in hopes of a comeback on bumper crops. When prices fell yet further, the suitcase farmers bailed out. They left huge swaths of eight states bare—and left the farm families who remained to face an unprecedented ecological disaster.

During the spring and summer of 1930, drought settled on the eastern United States, shattering all previous records. Along the Mississippi River that had swelled to unimaginable heights just three years earlier, river captains wondered whether their barges would be able to make it south to New Orleans.

As the drought moved across the rest of America, its center shifted from the East to the Great Plains beginning in 1931. The Dakotas became as arid as the Sonoran Desert. As the rain dried up, intense heat set in. The mercury first climbed to 115 degrees, then to 118. In summer 1934, thermostats in Illinois pushed over 100 degrees for so long that 370 people died. Two summers later, *Newsweek* described the country as a "vast, simmering caldron." More than 4,500 died from extreme heat.

Grasshoppers swarmed in buzzing black clouds like the plagues of Exodus. After gnawing the last of the anemic wheat stalks and corn nubs in the fields, they began to devour fence posts and the wash hanging on clotheslines. Farmers lost their crops to the drought, and their land, farm equipment, and homes to the banks. By 1936, farm losses reached $25 million a day. More than two million farmers were drawing relief checks.

Combined with the barren landscape, the rain-free years of the 1930s brought worse than the heartbreak and hunger of Uriah's generation. When the swift old prairie winds met the newly stripped ground, they kicked up a novel type of storm, the likes of which had never been seen. These violent storms really did follow the plow. Rather than rain, they carried millions of pounds of dirt.

The winds that nearly blew Uriah off his sod house were back, sometimes howling through at sixty miles an hour or more. With neither the prairie grass nor the wheat crops to stand against the gales and anchor the drying soil, farmers lost the earth itself. As soils calcified, dirt started to blow off the land, turning daytime to darkness. John Riley "J.R." Davison, born in 1927 in a tiny Oklahoma panhandle crossroads called Texhoma, was haunted by his boyhood memories of black blizzards that blocked the sun. "We watched that thing and it got closer and seemed to kindly grow. It was gettin' closer. The ends of it would seem to sweep around. And you felt like you were surrounded. Finally, it'd just close in on you, shut off all light. You couldn't see a thing. The first one or two that happened, people thought the end of the world had come. Scared 'em to death. Travelers comin' down the highway didn't know what to do. They were just hysterical."

Sometimes, the black clouds would boil tens of thousands of feet into the sky like angry cumulonimbus, then rumble cross-country on high-altitude winds. In 1934, drought parched forty-six of the forty-eight states. A massive front that May picked up 350 million tons of earth as it swept over the plains, carried it east, and dumped 12 million pounds on Chicago. The dry storm next darkened the skies and dirtied the air in Boston, New York, Washington, and Atlanta, before moving over the sea. Ships three hundred miles out in the Atlantic Ocean reported a quarter inch of dust on their decks.

As farmers watched dirt bury their fence posts and wagons, they tried desperately to keep it out of their homes, their drinking-water wells, and the lungs of their children. Babies and the elderly were most vulnerable to an epidemic of respiratory infections including "dust pneumonia." The number of people who died in the Dust Bowl is unknown, but with heat deaths and pneumonia combined it was in the thousands. More than 250,000 other souls packed up all they could carry and fled, like the Joad family in John Steinbeck's *The Grapes of Wrath*.

The worst single day came on April 14, 1935, known as Black Sunday. A sunny day had finally dawned, drawing many people outside for picnics and Sunday drives. Midafternoon, the warm air temperatures plummeted—falling by as much as 50 degrees in a few hours. A monstrous black blizzard appeared on the northern horizon and moved in with silence and speed. Woody Guthrie was twenty-three years old and living in Pampa, Texas. He watched the "freak-looking thing" bear down on the town "like the Red Sea closing in on the Israel children." He later recounted how the day inspired him to write the song "So Long, It's Been Good to Know You":

> It got so black when that thing hit, we all run into the house, and all the neighbors congregated in the different houses around over the neighborhood. We sit there in a little old room, and it got so dark that you couldn't see your hand before your face, you couldn't see anybody in the room. You could turn on an electric light bulb, a good, strong

electric light bulb in a little room and that electric light bulb hanging
in the room looked just about like a cigarette burning. And that was
all the light that you could get out of it.

    A lot of the people in the crowd that was religious-minded, and
they was up pretty well on their scriptures, and they said, "Well
boys, girls, friends, and relatives, this is the end. This is the end of
the world." And everybody just said, "Well, so long, it's been good to
know you."

The Great Mississippi River Flood of 1927 and America's Dust
Bowl, which began just three years later, were inverse disasters, triggered by hard rains and hardly any. But they shared much in common,
not least of which a bitter national wake-up from the dream of Manifest Destiny. The Jeffersonian vision of small yeoman farmers living in
idyllic harmony with the land could not be realized out of harmony with
the rain. Those hardest hit by the Dust Bowl were the tenant farmers
who fled, and the small family farmers who stayed behind. Hardest hit
by the Mississippi flood were black sharecroppers in the Delta, leading
more blues to be recorded on the subjects of heavy rainfall and inunda-
tion than any other natural phenomenon. Before the floodwaters had
receded, Bessie Smith was out with "Back-Water Blues" and "Muddy
Water (A Mississippi Moan)," while "Blind" Lemon Jefferson recorded
his album *Rising High Water Blues*.

    People, since its raining, it has been for nights and days
    People, since its raining, has been for nights and days
    Thousands people stands on the hill, looking down where they used
    to stay.

After Uriah Oblinger found care for his three motherless girls, he
sold his possessions and left Nebraska for Minnesota to be near family
and find work—and hopefully, a wife. He married within about a year,
brought home Ella, Stella, and Maggie, and went on to have three
more daughters. Another son died in infancy.

For a while, Uriah returned to his tumbleweed ways, working for a railroad surveying company. But he missed his family. He tried taking everyone back to Fillmore County, Nebraska. His farm failed. He heard good things about Kansas. He did not succeed there. He headed to Missouri. Another farm failed.

In fact, he never again found the success he'd had with Mattie in their seven rainy years together. His final letters contain no rainbows. They recount the failing national economy; mounting personal debt and collectors calling; and suggestions to his wife, now back in Minnesota, about what additional possessions she could sell for cash: Uriah's workbench, his planes, his pitchforks, three wagon box rods, the grain cradle, the rake.

In the end, he returned to Nebraska, where his firstborn, Ella, and her husband were making a go. His last surviving letter was to one of Mattie's brothers: "Our crops here on the divide were almost a total failure again," he wrote.

"We had no rain."

EIGHT

THE RAINMAKERS

I n August 1891, Robert St. George Dyrenforth, a Washington patent attorney, arrived by train at the small Midland, Texas, station in a desolate stretch of the southern plains. He had sent ahead a freight car with a bewildering assemblage of rabble: mortars, casks, barometers, electrical conductors, seven tons of cast-iron borings, six kegs of blasting powder, eight tons of sulfuric acid, one ton of potash, five hundred pounds of manganese oxide, an apparatus for making oxygen and another for hydrogen, ten- and twenty-foot-tall muslin balloons, and supplies for building enormous kites.

Dyrenforth, his odd freight, and a small group of experts, "all of whom know a great deal, some of them having become bald-headed in their earnest search for theoretical knowledge," joked a pundit, were met by local cattle ranchers. On mule-drawn wagons, they trekked twenty miles northwest to one of the largest cattle operations in the country, owned by a Chicago meatpacking king and spread over four arid counties.

Sporting the pith helmets and knee-high hunting boots of a Teddy Roosevelt safari, the out-of-towners struck a blushing contrast with the cattlemen in their cowboy hats and boots. But the city boys and cowboys worked together over the next week as they followed Dyrenforth's instructions for setting up what can only be described as a series of battle lines.

Along the front line, Dyrenforth erected sixty homemade mortars about fifty yards apart. Mounted to the ground atop wagon-axle boxes, the mortars pointed to the sky at 45-degree angles. Also on this first line, Dyrenforth "made several mines or blasts by putting sticks of dynamite and rackarock into prairie-dog and badger holes," he reported.

A half mile behind the first, Dyrenforth set up a second line consisting entirely of homemade electrical kites, each about as big as a family dinner table. He staggered the kites between the mortars. Without enough men to fly them, he tied them to "the mesquite bushes, the chapparal bushes, and the catsclaw bushes."

On a gently sloping prairie about a half mile behind the kites, Dyrenforth set up a final line with ten- and twenty-foot-tall balloons and the hydrogen apparatus to inflate them. A photograph from the scene shows one of the pith-hat men filling a comically large balloon in the tall prairie grass, a windmill behind him. The black-and-white image conjures Dorothy's return to Kansas and looks every bit as imagined.

But Dyrenforth was real, and so was his scheme: He planned to bomb the hell out of the skies over West Texas to try to make them rain. Not only that, it was all being bankrolled by the United States Congress.

When the Erie Canal opened in 1825, New Yorkers felt so triumphant that they threw the biggest series of celebrations since the Declaration of Independence. In cities and towns up and down the canal, thousands and thousands of people streamed to waterfront parties and parades or boarded festively decorated boats to join a flotilla from Buffalo to Albany and down the Hudson River to New York City. "The delight, nay, enthusiasm, of the people was at its height," gushed one correspondent. "Such an animating, bright, beauteous, and glorious spectacle had never been seen." And such an explosive cacophony had never been heard. The statewide celebration began on the sunny Wednesday morning of October 26. As the steamboat *Seneca Chief* launched into the canal at Buffalo, a gunner set off the first ear-splitting blast in a lengthy salute of 32-pound cannons. The cannons had been set every ten miles along the nearly four-hundred-mile route. When the men at the second cannon heard the crack of the first, they would fire, and so on down the line all the way to Sandy Hook on the Atlantic Ocean.

Ringing out with the cannon booms were the ceremonial discharges of every gun that could be rounded up from the Battle of Lake Erie, a decisive naval victory in the War of 1812. At each town the flotilla passed, locals also set off "deafening fireworks" and artillery companies fired their own salutes. The smallest villages were determined to create celebrations as noisy and elaborate as any in the larger towns. Atop the cannonading, every town rang its church bells and held a parade with a marching band. Every canal-front party and every boat deck likewise had a band, and many families shot off their own firecrackers and muskets.

All day and all night, the air cracked with a racket to kill the weak-hearted and call up the dead. And when the celebratory blue skies of Wednesday turned to cold, gray dampers that washed out the Thursday parades, opinion spread that the noise had been loud enough to call the rain, too.

The belief that thundering noise could bring rain or vanquish it was as old as, well, thunder, lightning, and rain. In Europe during the Middle Ages, church bell ringers also had to scout for the severe storms that plagued those grim times. Upon sight of a black thunder-cloud, they would head to the steeple to set the bells clanging—not to warn villagers, but to try to break the storm. (Clutching the end of a wet rope hung from the tallest point in the village in the days before lightning rods was not the wisest way to ride out a thunderstorm. Prior to the Enlightenment, it was not uncommon for church bell ringers to be killed by lightning.)

From the time of Plutarch, rainfall also had been associated with

war; the Greek essayist wrote that "extraordinary rains pretty generally fall after great battles; whether it be that some divine power thus
washes and cleanses the polluted earth with showers from above, or
that moist and heavy evaporations, steaming forth from the blood and
corruption, thicken the air." By the eighteenth century, storied connections between war and rain had become linked to the blasts of heavy
gun and artillery fire—linked not by scientists, but by military leaders
including Napoleon.

n 1842, the U.S. government had hired its first official meteorologist, the brilliant James Pollard Espy, widely admired at home and in
Europe. Espy had championed the national system of weather observers, "the only information now wanted to predict rain." His convective
theory of rainfall, which explained how storms are driven by warm,
humid air rising in a column, was far ahead of its time. It earned him
the Magellanic Prize of the American Philosophical Society and his
nickname, "Storm King."

But the stumble in his otherwise esteemed career was his intellectual leap to rainmaking by fire. While many people (not, for the most
part, scientists) believed the loud concussions of artillery caused rain
following battles and even Fourth of July celebrations, Espy thought it
was the convective heat. He became convinced that cutting and burning huge tracts of forest would bring quenching storms to the arid
regions. He proposed that the government maintain gigantic timber
lots in a belt from the Great Lakes to the Gulf of Mexico along the
western frontier. When rain was needed, or even on a regular schedule,
some of the lots could be set ablaze. A long curtain of showers would
form and sweep eastward across the states to the seaboard, fulfilling
farmers' crops and dreams.

Espy's critics did not fear his plan would fail but that it would
succeed—placing the power of rainfall into the hands of the federal
government. The idea especially alarmed antebellum southerners. "He

might enshroud us in continual clouds, and indeed, falsify the promise that the earth should be no more submerged," argued a Kentucky senator, "and if he possesses the power of causing rain, he may also possess the power of withholding it."

Southern congressmen managed to block Espy's proposals for rain by controlled burn through the 1830s and '40s. For years, they would view him as a warning symbol of government control: "I would not trust such a power to this Congress," a South Carolina senator declared in the 1850s. Rain "is a power which none but God can rule with justice. As long as you leave it to the temptation of selfish man, it will go to make the rich richer and the poor poorer."

The desperate drought of the 1890s finally convinced Congress to invest in rainmaking experiments. The only redeeming influence of the decade-long drought was that it tamped down the "rain follows the plow" theory hard-sold by the railroads and other boosters. Clearly, the construction of rail and telegraph lines was not bringing rain to the plains. John Wesley Powell said as much in his *Arid Lands* report, and warned that settlement was changing the water cycle in other ways: Clear-cutting forests meant more runoff and less evaporation. Irrigation diminished flows to lakes. Digging drains meant drying marshes.

Powell's predictions for the hapless small farmers were beginning to come true, too. Nostalgically known as the Gay Nineties (Naughty Nineties in the U.K.), the 1890s were nothing but grim for Uriah Oblinger and the other farmers who'd swarmed into Kansas, Nebraska, and the Dakotas as young men to bust sod and fulfill the nation's destiny. Yet the belief that Americans would ultimately overcome the climate and even control it remained strong. Those who knew the least about the nascent science of meteorology seemed to believe it most fervently.

The first champion of the theory of rain by concussion was a celebrant at the Erie Canal parties in New York. J. C. Lewis "took note of a very copious rain that immediately followed the discharge of ordnance during the celebration of the meeting of the waters of Lake Erie and the Hudson." Writing nearly four decades later in the first year of the Civil War, Lewis pointed out that the day after the recent battle of Bull Run, rain on the battlefield "was copious all day and far into the night."

Centuries of rainy battlefields, his witnessing of artillery fire blasting rain from the skies at the Erie Canal, and the opening hostilities of the Civil War "fully established the fact," Lewis wrote in a 1861 letter printed in America's first scientific journal, *The American Journal of Science and Arts*: "The discharge of heavy artillery at contiguous points produces such a concussion that the vapor collects and falls generally in unusual quantities the same day or the day following."

The war slogging in the wet South did much to popularize the idea. Fighting in the godforsaken mud convinced hundreds of thousands of officers and soldiers, Union and Confederate alike, that gunfire brought down torrents. Governor Joshua Chamberlain of Maine, who volunteered for the Union Army and became a brigadier general, was one of many believers who saw with their own eyes how hard rains followed hard battles. Chamberlain noted that cleansing rains followed "Antietam, Fredericksburg, Chancellorsville, Gettysburg, the Wilderness, Spotsylvania, Bethesda Church, Petersburg, Five Forks," as well as small engagements with sharp, concentrated fire. "This fact was well noticed, and is well remembered by many a poor fellow who, like myself, has been left lying, desperately wounded, after such engagements—for these rains are balm to the fever and anguish of the poor body that is promoted to the ranks of 'casualties.'"

In 1871, a retired Civil War general and Chicago civil engineer named Edward Powers published his treatise *War and the Weather, or the Artificial Production of Rain*, in which he reviewed some two hundred battles to show that rain followed artillery barrages, usually within a day or two. Powers urged the government to "verify the truth of the theory and determine its limits and conditions." He suggested bringing two hundred siege guns of various calibers out of retirement from the federal arsenal at Rock Island, Illinois, for two experiments: one to see if a storm could be created in clear skies, and another to see if an approaching storm could be made to deviate from its natural course.

While Powers had considerable support from the public, particularly with so many Civil War veterans on his side, many scientists were dubious. Powers seemed hopelessly behind on what nineteenth-century science knew about the workings of the atmosphere and rainfall. Government meteorologists did not believe gunpowder explosion could bring rain. The war had been fought in the wettest region of the country; rain normally fell every few days.

But, then as now, Congress was less moved by its own scientists than by the influential uninformed—particularly some of the nation's major cattle ranchers suffering in drought. In 1890, Charles Benjamin Farwell, a U.S. senator from Illinois who owned extensive ranchlands in Texas, pushed through a bill that approved nine thousand dollars for a series of field experiments on rainmaking by concussion, to be conducted by the Department of Agriculture through its Division of Forestry. The chief of the division was the Prussian-born and -educated Bernhard Fernow, the first formally trained forester in the United States. Fernow had no faith in the project he'd been ordered to supervise. He complained that he had neither men nor means to explode ordnance into the skies, and that using the federal money as proposed by Congress "would hardly fail to be barren of results."

His complaints and ridicule got Fernow "excused from planning or conducting" the rainmaking experiments, no doubt a relief. The project was then delegated to the assistant ag secretary, who also squirmed out. He turned the entire project over to a special agent and gave him free rein. The agent was Robert St. George Dyrenforth, who had repeatedly shared his ideas about rainmaking by concussion with Senator Farwell.

This is how Dyrenforth ended up on the Texas prairie in August 1891 with his battery and balloons. "A patent lawyer in Washington who knows more about explosives and the manner of exploding them than any other man," observed Fernow, the forestry chief. "I strongly advise everybody to have his ark ready for the deluge."

The night of August 9, an aging General Powers, author of *War and the Weather*, joined Dyrenforth in Texas. The experiments had not yet begun, but the party shot off a number of ground charges for practice. The next day, rain. Someone telegraphed Senator Farwell, who in turn alerted the newspapers, which in turn jumped on the big news: "They Made Rain," said Denver's *Rocky Mountain News*. "Heavy rain fell, extending many miles," proclaimed the *Washington Post*. "Made the Heavens Leak," reported the New York *Sun*.

When the real experiments began, Dyrenforth had problems on all three of his lines. The homemade mortars failed to make much of a concussion, so the men exploded dynamite and rackarock from prairie-dog holes ("not much noise") or large, flat stones ("the concussive effect on the air was strong"). The balloons were unwieldy in the winds and the fuses too long; a ten-footer they managed to explode blew miles away before they heard the faint pop.

No rain followed the first balloon explosion. Still, Dyrenforth reported dark clouds were "seen to form in the west-southwest and rain fell from them heavily, accompanied by lightning," as if to credit the balloon, which had blown off in the exact opposite direction. His sixty-page report to Congress often tantalizingly describes "dew on the grass" or storm clouds on the horizon with visible streaks—giving the reader a soaking sense and himself indirect credit even when there was no rain.

For ten days beginning on the morning of August 17, the sky soldiers dynamited, bombed, and shot the atmosphere; flew their kites; and exploded their balloons, all on an erratic schedule that seemed based as much on how the men were feeling (many were sick from drinking alkaline well water) as on weather and wind conditions.

The ranch saw rainfall at various times over the ten days, and Dyrenforth reported each drop with jubilation. This was a rub for the federal meteorologist George Curtis, who'd been sent to the outpost to monitor the experiment, and for the only two reporters on the scene, one from the *Chicago Farm Implement News* and the other from *Dallas Farm and Ranch.*

Farm Implement News accurately pointed out that it was the rainy season for Midland. The North American monsoon sweeps into the region between mid-June and mid-September. Even in drought, western mid-Texas, New Mexico, and Arizona get some thunderstorms and rainfall at this time of year.

In the Midland rain-concussion trials, then others in El Paso, Corpus Christi, and San Diego, Texas, Dyrenforth and his team took credit for rain already predicted by the Weather Bureau, for showers that fell considerable distance from the sites, and even for rains that commenced before any explosions: The farm journalists sometimes witnessed the pith-hats shooting up rain-swept skies.

Unfortunately, most newspapers did not have reporters on the scene. Nor did they check the weather forecasts against the boosterish special reports issued by Dyrenforth's team. Instead, they repeated the hype and added their own. The *Washington Post* described the .02 inches of August 18 as a "hard rain" that fell immediately after an explosion and continued for four hours and twenty minutes. The New York *Sun* called it "a great success" that triggered six hours of rain over a thousand square miles. The newspaper predicted that "more than one Congressman will go to Washington this winter with a rain-making bill in his pocket."

The *Sun* got that much right. The following year, Congress again ignored the advice of federal scientists and approved ten thousand dollars for additional rainmaking experiments in 1892.

In the second year, the public and press began to see how rainmaking rhetoric differed from the rainfaking reality. The end of October 1892 found the team at Fort Meyer, across the Potomac from Washington, testing several new explosives. The nocturnal explosions rained nothing but "profanity in seventeen different languages drizzling down from chamber windows" as Washingtonians protested their disturbed slumber. After the Fort Meyer preliminaries, the rainmakers moved on to San Antonio, Texas. In November the expedition established Camp Farwell at Alamo Heights just north of the city. During the week of Thanksgiving, federal troops helped attack the skies. With little or no rain by early December, it was Dyrenforth who came under attack. "The scheme has gone up like a rocket and come down like a stick," opined the San Antonio *Evening Star.*

The cloudless skies in San Antonio brought an end to the government's nineteenth-century rainmaking experiments and stuck Dyrenforth with an irresistible nickname, "Dryhenceforth." When he wrote to U.S. agriculture secretary J. Sterling Morton to inquire about the $5,000 remaining in his budget for 1892, Morton replied that "we do not desire to cannonade the clouds any longer at government expense," and said the unspent balance would revert to the Treasury.

Curtis, the federal meteorologist who'd been assigned to observe the Midland experiments but whose critical report was never published, blasted the government's foray into rainmaking in a series of articles and a blunt letter to Fernow. The American public had been misled, he said, and the federal government was partly to blame. Experiments with no method, under management of one man, had been passed off as credible science. The fact that they were carried out in the name of the government, and reported as successful by so many newspapers even though they failed, gave the public a false confidence in rainmaking, Curtis warned. That left people particularly vulnerable to charlatans, who now appeared throughout the dry regions as if from the clouds themselves.

Working in explosives, chemicals, and gases, mostly in secret, late nineteenth-century rainmakers—often known as "smell-makers" for their rank ingredients—were busiest on the long-suffering Great Plains. On balance, there seemed to be as many well-intentioned efforts as dishonest ones. Throughout Kansas and Nebraska, governments purchased tons of dynamite to try to blast down rain. In Nebraska's northwestern panhandle, farmers formed the "Rain God Association" and raised $1,000 for gunpowder. On a hot July day, they set up a 250-mile-long cannonade of "Rain God Stations" on high peaks. They discharged all the powder at a prearranged second. No rain followed.

The era gave rise to the traveling rain man. One of the best known, Frank Melbourne, would turn out the biggest charlatan, too. Called "the Rain Wizard," "the Australian," or "the Irish Rainmaker," Melbourne was born in Ireland and had lived in Australia, where he was said to have honed his rainmaking skills. He claimed he'd been forced to flee to avoid retribution for conjuring floods.

Tall and dark-bearded, Melbourne showed up in 1891 in Canton, Ohio, where his brothers were living. His rainmaking demonstrations soon became a popular public spectacle. He took bets on whether he could make it rain, fattening his purse more often than not.

Melbourne began to charge $500 for a "good rain," one that would reach from fifty to a hundred miles in all directions. Unlike the ordnance exploders, his methods were quiet and mysterious. His "Rain Mill" involved a crank and gases, but he never let anyone see it. He toted it in plain black gripsacks, along with a large revolver "to discourage too curious spectators." In Canton, he carried out his demonstrations in a shed, with a brother posted outside to record the bets and keep anyone from peeking in.

Melbourne came to Cheyenne, Wyoming, when twenty-three farmers pooled their money to buy a rainy day. He disappeared into a stable, covered the windows with blankets, and stuffed the cracks for total secrecy. The next day brought violent thunderstorms and the heaviest rains of the year. Farmers cheered, and so did newspaper editors: "The irrigation ditch can go," proclaimed the *Cheyenne Daily Sun*. Excitable press inflated Melbourne's reputation even as many of his attempts failed to squeeze a drop from the sky. He drew no showers for Goodland, Kansas, despite several days' wizardry. It rained in other parts of the state, though, and he said the wind had blown his work off course. The *Chicago Tribune* played it like this: "Melbourne Causes the Rain to Fall. Complete Success Attends His Latest Experiments at Goodland." It was no wonder Melbourne grew more and more popular and charged higher and higher fees, which may have helped do him in.

After a string of failures, someone figured out that the dates Melbourne selected for rainmaking were identical to those for which rain was forecast in the popular weather almanac of the day, the quirky *Rev. Irl R. Hicks Almanac*. In his final swirl of mystery, Melbourne was found dead in a Denver motel room in 1894. Police ruled it a suicide. But his rainmaking legacy lived on in the Great Plains. Melbourne had given Goodland a final squeeze by selling his secret formula and copies of his rainmaking machines to local businessmen.

After Melbourne left Goodland, three new companies—the Inter-State Artificial Rain Company, the Swisher Rain Company, and the Goodland Artificial Rain Company—began selling the rain throughout Kansas and beyond. A fourth Goodland rainmaker, Clayton B. Jewell, was the town's young rail dispatcher. He said he'd figured out Melbourne's method. The Rock Island Railway bankrolled Jewell's experiments, fixing him up in his own rainmaking rail car with gas pipes sticking out the roof. Offering his services to thirsty communities and farmers for free, he was a popular figure on the plains, where he claimed credit for sixty-six rains. In the summer of 1899, Jewell struck out on his own for Los Angeles, where he had been invited to work his magic to help lift a drought. He coaxed the Californian clouds for sixty hours straight, and got nothing.

A few dry years dried up Jewell's reputation, as it did the lesser-known rainmakers. (Here revealing the irony of the legal profession's term "rainmaker" for those well-connected lawyers in the firm who bring in new clients and their money.)

As the century turned, good rains returned—along with a new rainmaker to dazzle Californians. Charles Hatfield would go down in history as the greatest rainmaker, or faker, of them all.

———

Born aptly in the rainmaking motherland of Kansas, Charles Mallory Hatfield's family moved to Southern California when he was a boy. He shined in his first jobs, as a sewing machine salesman, then manager of the Home Sewing Machine Company of Los Angeles. But his passion lay in the machine of the atmosphere. Hatfield read and reread classic weather texts of the day such as *Elementary Meteorology* by William Morris Davis. The San Diego Public Library still has his well-thumbed, underlined copies. By 1902, he was dabbling in rainmaking at his father's ranch outside San Diego.

Hatfield said the swirling steam of a teakettle gave him the idea to climb a windmill at the ranch, heat some chemicals in a pan, and send the vapors ambling into the sky. When a heavy storm descended after his first try, Hatfield was convinced that he either had figured out the recipe for rain—or could make people believe he had. He got into professional rainmaking on a bet, when he claimed he could draw eighteen inches to Los Angeles for the winter and spring of 1904–1905. "The regular fall rarely exceeds eight or ten inches," he added. Having studied his books and Weather Bureau reports, he likely knew better. L.A.'s annual average rainfall is fifteen inches. In times of drought, including the year Jewell was called west in 1899, the city can see as little as five. During what we now know as the El Niño cycle, L.A. can draw closer to twenty. Ending an unusually dry decade, the winter of Hatfield's bet was such a year.

Hatfield had not only El Niño on his side, but trim good looks and a modesty that appealed to businessmen and civic leaders. "I cannot make it rain," he would say. "I simply attract clouds, and they do the rest." Thirty prominent Los Angeles businessmen signed on to give Hatfield $1,000 if he could draw eighteen inches by May. Hatfield erected a derrick twelve feet high in the San Gabriel Mountains above Pasadena. He set to work heating and mixing his chemicals and sending them aloft. He explained his technique of "subtle attraction" to a reporter for the *Los Angeles Examiner*:

When it comes to my knowledge that there is a moisture-laden atmosphere hovering, say, over the Pacific, I immediately begin to attract the atmosphere with the assistance of my chemicals, basing my efforts on the scientific principle of cohesion. I do not fight Nature as Dyrenforth, Jewell and several others have done by means of dynamite bombs and other explosives. I woo her by means of this subtle attraction.

After a dry Rose Bowl Parade on New Year's as requested by the locals, the Los Angeles skies rained furiously that winter. The last inch specified in Hatfield's contract fell a month before his deadline. The triumph created demand for Hatfield and his rain derricks throughout the West Coast, including Canada and Mexico. He was soon credited with filling the streams that kept local mines in operation in Grass Valley, California; replenishing the reservoirs of the Yuba City Water Company; and bringing much-needed rains to the parched badlands near Carlsbad Caverns.

Hatfield discouraged crowds from congregating around his towers. But those who caught a glimpse were mesmerized. Fuming chemicals billowed and swirled in great plumes above the derricks, then disappeared into the air. Hatfield worked like a madman, clambering across the scaffolds, pouring his concoctions, and "keeping the mixture moving" as it evaporated.

Critics pointed out that Hatfield's contracts gave him almost unbeatable odds. He always worked during rainy season, and his contracts extended over a hundred-mile radius, "which increased his chances of apparent success a hundredfold," writes the science and technology historian Jim Fleming. His most dogged detractors were the scientists at the U.S. Weather Bureau. Each time a newspaper lauded Hatfield, the bureau chief Willis Moore sent in scathing rebuttals from Washington, chasing Hatfield around the nation in print. In Hatfield's home region in San Diego, the local weather bureau commissioner wanted him prosecuted for fraud.

But the scientists again lost the battle for the public mind and heart.

"As far as many townsfolk were concerned, this was an ordinary fellow, with no fancy degrees behind his name, whose woodshed brew delivered what lofty scientists and their delicate instruments could not," writes the weather scribe Nick D'Alto. Country newspaper editors agitated for the government to buy Hatfield's formula "and either prove it to be worthless or apply its merits."

Fittingly, a burst of storms would cinch Hatfield's place in history. In San Diego, 1915 marked the fourth year of a devastating drought. With reservoirs only a third full, the city council agreed to a proposal from Hatfield: Over the following year, he would make enough rain to overflow Morena Reservoir. If Morena topped its banks by December 1916, the city would pay him $10,000. If not, he would be owed nothing.

In January, Hatfield got to work. He'd been atop his derricks brewing his secret blend of chemicals just a few days when the skies began to drizzle. The drizzles turned to steady rains. The rains turned to record torrents; more than 28 inches fell that month. Morena Reservoir overflowed. On January 27 the Lower Otay Dam burst, sending a wall of water into downtown San Diego that killed dozens of people, destroyed many more homes, and washed out all but two of the city's 112 bridges.

The disaster became known as "Hatfield's Flood." Armed vigilantes were said to have gone after him and his brother, who fled on horseback. They returned the first week of February and held a press conference, Hatfield striking the demeanor "of the proverbial conquering hero, home from the fray and awaiting the laurel wreath," wrote the *San Diego Union*. He announced he had fulfilled his contract to overflow Morena Reservoir, and now expected his $10,000. The shrewd city attorney argued the deluge was "an act of God." Hatfield ultimately filed a lawsuit. The city lawyers said San Diego would pay only if he would sign a statement assuming responsibility for the flood and relieving the city of damages; some $3.5 million.

The suit dragged on until 1938. Hatfield never received his fee.

Surely the publicity was worth more. For the next fifteen years, Hatfield built his derricks around the West and on a few international assignments. In 1922, the *New York Times* ran a story about his trip to Italy to help break a drought: "He was anxious to explain his secret process to Pope Pius, and if the Pontiff agreed he would try to induce rain to fall on the Vatican gardens."

He made a trip to Honduras to save seven hundred acres of drought-stricken banana lands for the Standard Steamship and Fruit Company of New Orleans. In Medicine Hat, Alberta, the United Agricultural Association paid him $8,000 for nearly five inches of rain that fell in the summer of 1924. Hatfield built his tallest tower ever, lived in a "roomy cook car" next door, and conjured his chemical brew up to eight times a day. Sightseers flocked to see, and a moviemaker to film. When the spectacle was over, some observers pointed out that Medicine Hat's long-term average rainfall in the three-month contract period was about six inches; Hatfield had cashed in on rains an inch below normal.

Even the greatest rainmaker of all time could not survive the 1930s economic collapse and Dust Bowl. Hatfield's fans urged President Franklin Roosevelt to bring him to the plains; six or eight of his derricks, they claimed, would bring an end to the disaster. But the era of celebrity rainmaking was over on all but stage and screen. Hatfield was the inspiration for Richard Nash's play *The Rainmaker*, later made into a movie starring Burt Lancaster and Katharine Hepburn.

Hatfield was eighty-two when he died quietly at his home in Pearblossom, California, in January 1958. It was a rainy El Niño winter. No one picked up the news until four months later, when the city of San Diego tried to reach him—perhaps for a history project or a documentary. The headline in the *Washington Post* read: "Charles Hatfield, the Rainmaker, Dies in Obscurity."

The progressives in government were long embarrassed about their inability to counter rainmaking, which had been born with the aid of the U.S. Department of Agriculture. When the ag secretary Jeremiah Rusk stepped down in 1893 to become governor of Wisconsin, he proclaimed rainmaking would go down "among the curiosities of so-called scientific investigation, in company with its twin absurdity, the flying machine."

In the twentieth century, both absurdities would move from the realms of quackery to those of science and engineering.

While working one day at McCook Field just north of downtown Dayton, Ohio, in 1922, Orville Wright heard an unfamiliar airplane engine overhead. He walked to his window overlooking the thousand-foot airstrip banked by wooden hangars and engineering labs, established by the army four years before as America's first base for experimental aeronautical research.

It had been twenty years since the younger Wright, dashingly adorned with a waxed black mustache, soared into history in the world's first lasting—all twelve seconds' worth—powered flight along the beach at Kitty Hawk. He and his older brother, Wilbur, had gone on to log hundreds of flights, win a patent war for their Flying Machine, establish a school for pilots, manufacture airplanes, and sell the U.S. government its first.

By this point, Wilbur had been dead ten years. Orville's mustache was flecked with gray and pruned of its gravity-defying curls. The man no longer challenged gravity, either. Wright had piloted his last flight and sold his airplane company for a reported million and a half dollars so he could devote his time to inventing. Then generating more patents than any other city in America, Dayton had become the nation's sanctuary for invention. Wright was its spiritual leader.

Wright had been a key aeronautical adviser to the U.S. military throughout World War I, and still kept an office at McCook Field, "the bridge over which homespun flying machines stepped into the realm of truly engineered aircraft," writes the air historian Peter J. Jakab. Air servicemen there perfected altimeters and airspeed indicators, earth-induction compasses and engine tachometers. McCook pilots finessed free-fall parachuting. They experimented with vertical flight. They completed the first nonstop coast-to-coast flight across America, in 1923.

Academic and industry scientists who could gain permission from Air Service brass descended on McCook as well. Agricultural researchers bombed nearby farm acreage with insecticides dropped from a powerful new weapon against grasshoppers, the crop duster. Meteorologists, for the first time, could hoist their instruments directly into the clouds.

It was this sort of nonmilitary experiment, on a summer day in 1922, that caught the ear of Orville Wright—and then drew his bewildered eyes.

The sky that June day was dotted with dense, white cumulus clouds. Wright peered through his window just in time to watch a small plane, a French-built Le Père, disappear into one of them.

About ten to fifteen seconds later, the plane emerged from the other side of the cloud, trailing what appeared to be smoke. On closer observation, Wright realized the long plume behind the plane was some sort of dust. The plane turned and dragged its dirt banner through the cloud again, and then again, for a total of five or six passes.

"The cloud began to fade away," Wright told a reporter, "and at the end of three or four minutes had practically disappeared."

The Le Père then nosed into another cloud, to the left of the first. After buzzing in and out of it several times, this cloud, too, began to fade. Then, the busy little plane tackled a third. "Within about ten minutes from the time I first saw the plane," Wright marveled, "three clouds had entirely disappeared."

America's aviation pioneer had witnessed man's first successful effort to control the atmosphere. Now that modern humans could fly like the gods of mythology, could they also make it rain like Jupiter Pluvius?

———

The McCook Field cloud experiments took flight in an odd partnership between an entrepreneur named Luke Francis Warren and the Cornell University chemist Wilder Bancroft. Grandson of the well-known historian, diplomat, and U.S. navy secretary George Bancroft, the rotund scion once called himself "a specialist in unorthodox ideas." Bancroft had a particular interest in colloid chemistry, which he defined as "the chemistry of bubbles, drops, grains, filaments and films." He pondered the mechanics of raindrops—and their positive and negative charges—in his book *Applied Colloid Chemistry*. It is not clear how he met Warren, who claimed a diverse background of adventures from London to South Africa to the San Joaquin Valley of California, "where he had a bitter personal experience with drought."

Warren dreamed of earning his fortune with the first legitimate rainmaking company. Bancroft put up the scientific credibility and considerable capital. Their theory was that spraying clouds with electrified sand could coax down rain. Droplets of water suspended in clouds were charged either negatively or positively, but a film of condensed air around each prevented them from coming together to make the larger drops that form rain. Infused with sand of an opposite electrical charge, the droplets would coalesce and fall. The era of flight made it possible, Bancroft said. He wrote to Warren in 1920, "It would probably be absolutely prohibitive in cost to produce rain by spraying clouds from beneath; but it is quite possible that you can get satisfactory results by spraying from above."

Bancroft and Warren put considerable effort into lobbying the military for aircraft for their trials. Bancroft's family connections must have been invaluable, and Warren wined and dined the contacts in Washington. It paid off. The U.S. Army Air Service provided planes, pilots, other staff assistance, and funding for initial field tests. The team contracted with the Harvard physicist E. Leon Chaffee to design equipment to charge sand particles and release them from the plane. The sand shot out the bottom from an electrified chute and scattered through the target cloud like a little dust storm.

Like Wright, many who witnessed the trials came away with a sense of dramatic possibility even though they saw no rain. The head of McCook Field, Major Thurman H. Bane, told the *Dayton Daily News* a few months after the first tests that the military would soon be able to "push clouds out of the way when they are interfering with aeronautics," and that "the Air Service will be able to make it rain whenever a deluge will be effective in combatting the enemy." Commander Karl F. Smith of the U.S. Navy came to observe, and remarked that seeing a cloud sliced up was "absolutely uncanny." The military applications were "so important," he wrote, that after a few more trials, the U.S. Navy Bureau of Aeronautics should try to buy rights to the technology. Warren and Bancroft might have cashed in, but instead they formed the A.R. Company, for "Artificial Rain," and applied for a U.S. patent for "Condensing, Coalescing, and Precipitating Atmospheric Moisture."

The U.S. Weather Bureau remained ever skeptical and deployed press releases just as it had battling Charles Hatfield earlier in the century. Dr. William Jackson Humphreys, the bureau's meteorological physicist, placed Bancroft and Warren squarely with the nineteenth-century rain fakers: "The idea of the college professor and his aviator friends out in Cleveland, to sprinkle electrically charged sand on a cloud while above it in an airplane, is picturesque and plausible," Humphreys wrote in a press release, "but won't work in commercial quantities."

"The Bureau lumps the work of Dr. Bancroft and Mr. Warren with a series of valueless devices, pronouncing all rainmaking schemes to be 'entirely futile,'" wrote the *New York Times*. When the newspaper asked Bancroft to respond, he telegraphed: "No use arguing with Weather Bureau. Prefer to wait for results and let them do the explaining." But there was little to explain. The military remained interested for a few more years, and field trials moved to the Aberdeen Proving Grounds and then Bolling Air Force Base in Washington. While the electrified sand could "knock the stuffing out of clouds," as Warren once put it, the project was essentially "attempting to do a big job with little money" and never managed to make rain or disperse fog, its other goal. With little to show after nearly ten years, more than twenty thousand dollars of his own money invested, and seven thousand more coaxed out of friends, Bancroft himself seemed to lose faith, too.

Still, McCook deserves credit as site of the first successful attempt to control the atmosphere. It did not exactly get its historic due. The old airfield's cramped proximity to downtown Dayton and too-short runway cut short its usefulness to the military, too. In 1927, McCook's missions were relocated to Wright Field, about ten miles east of Orville's old bicycle shop, now part of the Wright-Patterson Air Force Base. Servicemen tore down the wooden buildings. They dismantled McCook's 97-foot wind tunnel, which had been crafted with elegant cedar rings. They broke down the propeller test rigs. They hauled the portable dynamometer and other equipment to the new location.

Today, McCook's place in aviation history is largely forgotten by Dayton residents, who may know the name only as the city's oldest strip mall, near the original field and now home to a bowling alley and peep-show theater. McCook Shopping Center is set back in a sea of old asphalt that is bowed and pocked by years of burying snow and brutal sun. Precipitation that falls on the pavement has nowhere to go, so it pools in large, oily depressions—perhaps the only rain ever harnessed at McCook.

In the 1940s, clouds of war ramped up the military's interest in cloud physics. This included aircraft icing, gas mask filters, and screening smokes, which the Germans used in '41 to hide the battleship *Bismarck* in the fjords of Norway. At the General Electric Research Laboratory in Schenectady, New York, Irving Langmuir, a Nobel laureate in chemistry, and members of his research team, Vincent Schaefer and Bernard Vonnegut, had a smoke-research contract with the military to improve gas masks and hide large areas from air raids. (Bernard's brother, Kurt Vonnegut, worked at the Schenectady lab, too, interviewing scientists and writing press releases about their work.) The War Department asked the team to look into the dangerous loss of radio contact with planes flying through electrical disturbances in snowstorms, known as precipitation static. Langmuir and Schaefer decided to base their investigations at the stormy, cloud-shrouded Mount Washington Observatory in New Hampshire.

As the GE scientists worked, they became fascinated with more basic cloud questions, such as why not all "supercooled" clouds—those containing liquid water droplets with temperatures below freezing—make snow. Serendipity helped figure it out. On a warm July day, Schaefer plunked some dry ice into a small freezer he was using as a cloud chamber to try to cool it down. Immediately, the cold cloud inside the chamber formed millions of tiny ice crystals. He removed the large chunk of dry ice and tossed in smaller and smaller ones, finding that even the tiniest grain would flood the chamber with ice crystals. It turns out that supercooled droplets need super-small bits, or nuclei, to cling to before they can make snowflakes—or raindrops.

When Schaefer shared his discovery with his boss, Langmuir "was just ecstatic and he was very excited and said, 'We've got to get into the atmosphere and see if we can do things with natural clouds.' So I immediately began to plan to seed a natural cloud."

On November 13, 1946, Schaefer rented an airplane, flew into a cold cloud over Mount Greylock in the nearby Berkshires, and dropped six pounds of dry ice pellets inside. The cloud "almost exploded" with ice crystals, Schaefer wrote in his notebook. Snow burst forth over the three-mile path of seeding. Langmuir, fifty miles away in the airport control tower, watched long, white streamers fall from the base of the cloud.

GE lost no time in asking the military for access to a better airplane that could fly into higher clouds—and in releasing the news to the world. The *New York Times* crowed the next day that "numerous practical applications were expected to result from the project, including storage of moisture in the winter for spring irrigation and water power programs, steering heavy snowfalls away from city areas, and providing snow for winter resorts." The *Times* quoted Langmuir as saying a pellet of dry ice the size of a pea could create enough nuclei for "several tons of snow" and "in a five-hour flight a single plane could generate hundreds of millions of tons of snow over a large area."

Letters, telegrams, and postcards poured in to GE and its scientists, full of ideas for commercial applications and those for public good. Movie producers wanted made-to-order blizzards. A search-and-rescue operation on Mount Rainier wanted to be able to clear clouds to spot downed aircraft. The Kansas State Chamber of Commerce wrote to President Harry Truman, asking the federal government to use GE's technology to combat drought. The ever-cautious U.S. Weather Bureau replied. Chief Francis Reichelderfer wrote that it was impossible to know how much precipitation had been coaxed and how much was natural. Besides, seeding worked only in special circumstances, and drought wasn't one of them. You cannot seed clouds in cloudless skies.

The next news flurries out of GE came in January 1947, after Bernard Vonnegut discovered that molecules of the compound silver iodide could also serve as artificial nuclei and "fool" cloud water droplets into crystallizing. Vonnegut experimented with shooting silver iodide from a generator on the summit of Mount Washington and "caused quite a nice snow squall downwind." The GE News Bureau made a considerable leap: Pumping silver iodide into huge masses of air, "it might be possible to alter the nature of the general cloud formation over the northern part of the United States during winter," the company declared in a press release. "It would prevent all ice storms, all storms of freezing rain, and icing conditions in clouds. . . . It should be possible to change the average temperature of some regions during winter months."

As the company's PR flacks and Langmuir touted large-scale weather control, another arm of GE tried to stop the bragging and even quash the research. The corporation's lawyers saw enormous risk in property damage and personal injury lawsuits. Manufactured snow and rain would lead, as surely as natural precipitation, to traffic accidents, injurious falls, floods, and who knows what unimagined mishaps. Plaintiffs couldn't sue over an act of God. They certainly would over an act of GE.

In February 1947, corporate bosses informed Langmuir's team that the company would transfer all cloud-seeding experimentation to the military. In what was later named Project Cirrus, the U.S. Army Signal Corps, Office of Naval Research, Air Force, and GE researched "cloud particles and cloud modifications." The contract was explicit that "the entire flight program shall be conducted by the government, using exclusively government personnel and equipment, and shall be under the exclusive control of such personnel." GE employees were restricted to laboratory work and reports.

Between 1947 and 1952, Project Cirrus conducted about 250 cloud experiments, from seeding to fire suppression. While the results of a few cloud-seeding runs were spectacular, trials in the clouds over Ohio in 1948 and California and the Gulf states in 1949 led researchers to conclude that seeding could neither initiate storms nor relieve drought. The best it could do was squeeze a little more precipitation from clouds that were going to rain or snow anyway. But Langmuir continued to see promise of pluviul proportions. After he credited seeding in New Mexico with a band of heavy rainfall across the United States, even some fellow scientists thought the Nobel Prize winner had lost his objectivity—or simply never understood the complexity of the atmosphere as well as he had his own field of chemistry. "No chemist, physicist, or mathematician who has not lived with and learned to understand this peculiar nature of meteorology can pass valid judgment on how the atmosphere will react if one interferes with the details of the natural processes," wrote the eminent meteorologist Sverre Petterssen in his memoirs as he pondered his work with Langmuir on cloud-seeding. "Moreover, to determine whether or not the atmosphere has responded to outside interference, it is necessary to predict what would have happened had it been left alone." From Robert St. George Dyrenforth to Irving Langmuir, that was always the rub: How could anyone be sure whether the clouds would have rained without the prompt?

While civilian scientists researched cloud-seeding as a way to combat drought in the arid West, hail on the plains, and hurricanes in the East, military strategists were chasing a darker, secret dream: weather as a weapon of war. In 1947, General George C. Kennedy, commander of the U.S. Strategic Air Command, declared, "The nation which first learns to plot the paths of air masses accurately and learns to control the time and place of precipitation will dominate the globe." Ten years later, an advisory committee reported to President Dwight Eisenhower that weather control could become a "more important weapon than the atomic bomb."

The U.S. military made its first large-scale attempt to unleash rain as a weapon during the Vietnam War. Beginning with trials in 1966, and continuing every rainy season until July 1972, "Project Popeye" dropped nearly fifty thousand loads of silver iodide or lead iodide in the clouds over Vietnam, Laos, and Cambodia to induce heavy rains. The idea was to flood out roads, cause landslides, and make transportation as difficult as possible well beyond monsoon season—essentially, to keep the Ho Chi Minh Trail a muddy mess and foil North Vietnam's ability to move supplies and personnel.

Operating out of Udorn Royal Air Base in Thailand without knowledge of the governments of Thailand, Laos, or South Vietnam, the Fifty-Fourth Weather Reconnaissance Squadron flew 2,600 sorties that dispersed 47,409 cloud-seeding flares—a payload code-named "Olive Oil"—to make rain over the trail.

In 1971, the nationally syndicated columnist Jack Anderson broke a story about the secret rain soldiers in the *Washington Post*. In 1972,

Seymour Hersh detailed the mission and the name "Project Popeye," revealed in the Pentagon Papers. The squadron dropped its last cloud-seeding flare a few days later. At congressional hearings convened by Senator Claiborne Pell of Rhode Island, the Nixon administration and military officials refused to admit the rain runs as they opposed Pell's resolution for a treaty to ban "any environmental or geophysical modification activity as a weapon of war."

The hearings revealed a great reluctance among many Americans to proceed with weather warfare. At the University of California at Los Angeles, the climatologist Gordon J. F. MacDonald warned that if rain could become a secret weapon, so too could floods, tidal waves, and droughts: "Such a 'secret war' need never be declared or even known by the affected population," MacDonald said. "It would go on for years with only the security forces involved being aware of it. The years of drought and storm would be attributed to unkindly nature and only after a nation was thoroughly drained would an armed takeover be attempted."

The House and Senate ultimately adopted anti-weather-warfare resolutions. In 1977, the United States, the Soviet Union, and other nations ratified a UN treaty prohibiting military "or any other hostile use" of environmental-modification techniques.

The results of Project Popeye have never been clear. Military scientists estimated rainfall increased between 15 and 30 percent over the Ho Chi Minh Trail. But was it the Weather Squadron's chemicals, or capricious monsoons? Project Popeye's final report claims "that judicious seeding of properly selected clouds resulted in remarkably increased cloud growth relative to the growth that would have occurred naturally." Today, leading atmospheric scientists as well as the intrepid private meteorologists who make their living seeding clouds say that would have been utterly impossible to know four decades ago. Even today, it is not entirely knowable.

One winter morning in Utah under bluebird skies with white-feather clouds, I drove south of Salt Lake City to the oldest commercial weather-modification company in the United States. North American Weather Consultants is tucked into a suburban office park behind a ballet studio and a chiropractor. The firm was founded in Southern California in 1950, part of an enthusiastic trend of private meteorology entrepreneurs who set up shop from the western United States to Australia in the wake of GE's cloud-seeding breakthroughs.

In the office, satellite cloud images swirl across large-screen televisions and a wooden cross hangs on the wall. The company's president, Don Griffith, and vice president Mark Solak are gray-haired meteorologists with soft arms in short-sleeved dress shirts. They are not men who could scramble up and down a rain derrick like Charles Hatfield. But like him, they stress humility in the cloud wizardry they perform for hydropower plants, ski resorts, and drought-ridden state and local governments.

Griffith has blown silver iodide into winter clouds over Utah's mountains for four decades to increase snowpack—and the annual runoff to rivers and reservoirs. North American Weather Consultants has about 150 ground-based generators set up in valleys and foothills on Utah's national forestlands. The devices look like tall space heaters arranged to warm the elk and snowshoe hares. When the meteorologists detect a winter storm with supercooled water droplets and optimal wind speeds for dispersing ersatz nuclei, they light the propane flames, which slowly burn the silver iodide into furling ribbons that seed the storm.

State water scientists in Utah estimate these efforts have increased the water stored in snow an average 13 percent annually, adding about 250,000 acre-feet a year to reservoirs and aquifers. But seeding remains controversial among many meteorologists, for both the results touted and the uncertainty of its influence on a system so dynamic, the father of chaos theory, Edward Lorenz, suggested it might be influenced by the flap of butterfly wings.

Indeed, after the unbridled excitement over GE's front-page head-lines in the 1940s, the next half century of weather modification was something of a letdown for meteorologists like Griffith and Solak who entered the field during a time of high interest and public investment. In the mid-1970s, the federal government was spending about $20 million a year on weather-modification research. But in the next couple of decades, that dwindled to nearly nothing. The chaotic nature of rain and snow, and scientists' inability to measure induced versus natural precipitation, meant the seeders could never definitively prove their results. It did not help that many of their rainmaking ancestors had been charlatans, their scientific forefathers exaggerators. A 2003 report from the National Research Council of the National Academies concluded: "There is still no convincing scientific proof of the efficacy of intentional weather-modification efforts."

Nine states have proof enough. As drought spreads and population grows in the arid West, cloud-seeding is one strategy among many balanced by water managers trying to keep reservoirs flowing. Since seeding is among the cheapest—a few dollars an acre-foot, compared with a few hundred dollars an acre-foot for reservoir storage or interbasin transfers, or a few thousand for seawater desalination—water managers see it as a small price to pay. Utah, California, Nevada, Idaho, Wyoming, Colorado, North Dakota, Kansas, and Texas all engineer clouds, as do many local governments, agricultural consortiums, and other rain-dependent businesses. A couple of hydroelectric companies in the United States and Australia have seeded clouds for more than fifty years without interruption.

Elsewhere, many foreign governments reached out to the clouds as the U.S. government stepped back from the science it helped invent. More than forty countries have active weather-modification projects, including Thailand, where the king himself—His Majesty King Bhumibol Adulyadej—holds a patent for "weather modification by royal rainmaking technology." Indonesia's disaster-mitigation agency has begun seeding monsoon storms over the Java Sea to try to make them rain before they hit Jakarta, where annual monsoon flooding can paralyze life and cause millions of dollars in damage.

China spends by far the most of any government on both actual seeding and research, with 47,700 employees on its weather-modification payroll, along with fifty cloud-seeding jets, 7,034 rocket launchers, and 6,902 mortars that look like Robert St. George Dyrenforth's wildest dreams come true. China claims that 560,000 cloud-seeding missions in the past ten years helped release nearly 500 billion tons of rain, or twelve times the water-storage capacity of the Three Gorges Dam, which spans the Yangtze River.

Roelof Bruintjes, lead scientist for weather modification at the National Center for Atmospheric Research in Boulder, Colorado, who also heads an expert team for the World Meteorological Organization, is skeptical of China's claims. But he says new models and technologies are tantalizingly close to clarifying the impacts of seeding. On a pair of mountain ranges in southern Wyoming, atmospheric scientists from the University of Wyoming, the National Center for Atmospheric Research, and the University of Colorado are wrapping up the most comprehensive seeding study in recent decades, including a blind statistical analysis of 150 seeding events over seven years.

Results were still nearly a year off when I talked to the lead researcher, a University of Wyoming professor named Bart Geerts. He told me the compounded evidence points to a general increase in precipitation on the seeded ranges; increases in the order of 10 to 15 percent could be attributed to seeding rather than to chance. But the study will not answer the other big question that has always followed the cloud-seeding industry: By taking a bit more rain and snow from the clouds, aren't they keeping it from the place where it would have fallen naturally?

Griffith of North American Weather Consultants characterizes the "robbing Peter of rainfall to water Paul" worry as a misconception; industry studies have shown that areas fifty to a hundred miles downwind from the seeding target also reap the benefits of increased precipitation. But Geerts told me the issue is much more complex in a global system where each rain or snow fall decreases the amount of water vapor downwind as it travels to the next mountain range and the next. Amid the mysterious workings of Earth and its atmosphere, one thing is clear—when we change one part of the rain cycle, we change another somewhere else.

In October 1947, the eighth tropical storm of the season spun out of the Caribbean, clipped western Cuba, and strengthened into a hurricane as it made landfall at Cape Sable in the Everglades, bringing spectacular thunderstorms and heavy rains to the southern tip of Florida. The deluges left two thousand people homeless in Dade County. Miamians rowed through the streets in skiffs and rafts to check on friends and tour the flooded city.

After speeding out of Florida and into the Atlantic Ocean, the Cape Sable Hurricane stalled 350 miles offshore from the Florida-Georgia border. Forecasters expected it to head farther out to sea. The GE and military scientists working on Project Cirrus decided it was a good time to try what the *New York Times* called their "daring meteorological experiment . . . to determine whether or not the colossal vortex that we call a hurricane can be broken by making it precipitate the thousands of tons of water that it contains."

On October 13, 1947, three Project Cirrus aircraft left MacDill Air Force Base in Tampa at 8:20 a.m. and arrived at the outer wall of the hurricane's eye a little before 11:00. Lieutenant Commander Daniel Rex of the navy observed an unusual, "exceedingly active squall line" of cumulonimbus clouds with tops boiling 60,000 feet in the sky, extending out from the center of the storm to the southeast. Rex decided to seed over a cloud shelf along the southwest side of the storm, dropping eighty pounds of dry ice, then an additional fifty pounds each into two convective towers. He described a "pronounced modification" of the seeded area, with overcast clouds turning to scattered snow clouds over three hundred square miles. For a day or so, forecasters were not sure what the hurricane would do. The evening of October 14, it turned back toward land.

The storm barreled into the southern Georgia coastline on the morning of October 15 at what's now called Tybee Island. Fishing trawlers sank and phone service was lost. Thousands of Savannah residents evacuated to churches and schools. "Savannah early this morning was hurriedly throwing up storm protections as the erratic Atlantic disturbance was moving almost directly toward the city," the local newspaper reported.

Still, Georgia averted major disaster and the U.S. mail was not interrupted, bearing out the sentiment chiseled in gray stone over the New York Post Office on Eighth Avenue: "Neither rain, nor sleet, nor gloom of night stays these couriers from the swift completion of their appointed rounds."*

Always on the side of big weather control, GE's Langmuir was certain that the seeding had changed the storm's direction and sent it toward Georgia. The federal government's chief hurricane forecaster in Miami, Grady Norton, believed it, too. But the Weather Bureau's official and final take, issued from Washington, was that the hurricane had already begun to change direction at the time of the seeding, and that a ridge of high pressure, not dry ice, turned the storm.

Soon after the fateful hurricane, GE announced it would no longer enforce its weather-modification patents. By putting them in the public domain, the company could be released from legal liability should its work be used to change the weather in some undesirable way.

Hurricane scientists were dismayed that the "unbelievably casual" seeding of the Cape Sable storm set back their modification research many, many years. Interest did not surge again for half a century, after

---

* The quote was a flourish by the building's architect, William Mitchell Kendall, who read Greek for pleasure. He found it in Herodotus's book of the Persian Wars, describing the Persians' loyal mounted postal couriers. Little did Kendall know Americans would make it the unofficial motto of the U.S. Postal Service.

Hurricane Katrina. In 2009, the billionaire Microsoft founder, Bill Gates, through the Seattle-based tech investment firm Intellectual Ventures, filed for patents with a handful of scientists working on technologies to weaken hurricanes—still chasing humanity's age-old dream to control the rain.

IV

—

CAPTURING

THE RAIN

NINE

WRITERS ON THE STORM

S teven Patrick Morrissey remembers his 1960s boyhood in Manchester, England, as sodden with rain and troubles. In his working-class neighborhood, the Victorian cobblestones never seemed to dry, rain sent black streaks down feeble windows, and mothers lined front parlors with buckets to catch the endless leaks. His primary school was a "bleak mausoleum" where "children tumble in soaked by rain, and thus they remain for the rest of the day—wet shoes and wet clothes moisten the air, for this is the way." It was fitting scenery for Morrissey's childhood isolation and sadness, which soon turned into clinical depression.

Manchester lies in the northwest of England, and its rainy renown, like Seattle's, is based more in cultural psyche than actual rainfall. Along with geography and grim skies, the two metros share similar creative contributions. It is perhaps no accident that the rain-famed cities of the United States and the U.K. birthed angst-filled independent rock genres: in Seattle, grunge; in Manchester, the moody indie pop of Morrissey's band the Smiths, along with Joy Division, New Order, and others.

In winter, drizzly Manchester gets an average half hour of sunlight a day. Its grit is born of history as the world's first industrial city, and that, too, had to do with the rain. The first steam-powered textile mill opened on Miller Street in 1781. Industry journals buzzed with the importance of high humidity for cotton manufacture; then and now, moist air meant fewer snapped threads. By the mid-nineteenth century, more than a hundred cotton factories—along with the one cranking out Mr. Macintosh's Waterproof Double Textures—huffed black smoke into Manchester's already ashen ceiling. Alexis de Tocqueville described their proud rise from the city's hilltops: "Huge enclosures give notice from afar of the centralisation of industry!" A young visitor named Friedrich Engels was more struck by the human enclosures. In horrified dispatches from worker neighborhoods, he described rain-rotting, filthy cottages built back to back along steep banks of the river Irk, which flowed black with raw sewage and the waste of the mills.

A century later, Manchester was bleeding population in a post-industrial economic collapse, pollution had all but killed the Irk, and working-class Mancunians were still living in the back-to-backs. One of them, a young guitar player named Johnny Marr, had grown up in the same sort of grim factory neighborhood as Morrissey. The two shared a taste and talent for dark lyrics, wit, and chords. Each kept hearing about the other, but Morrissey was too reclusive to seek out compatriots. It was Marr who came and banged on his door one day in the summer of 1982, wanting to write songs together.

The pair formed a band and called it the Smiths. Morrissey and Marr co-wrote songs that appealed to the sort of sensitive, alienated teens they'd been, in the words of one music journalist, "a vast company that loves misery." Marr said that the band's "usual default setting was Manchester in the rain." Their 1984 hit "William, It Was Really Nothing" (number 431 on *Rolling Stone*'s list of the 500 greatest songs of all time) began with the repeating phrase: "The rain falls hard on a humdrum town / This town has dragged you down."

The more famous he became, the more Morrissey bemoaned his youth in "Victorian, knife-plunging Manchester." But creatively, this town did anything but drag him down. "Teenage depression was the best thing that ever happened to me," he once told an interviewer, for it sent songs "sloshing around in my head." Their melancholy lyrics helped make him the most iconic British pop artist of his time, and even birthed a genre, mockingly dubbed "miserablism," beneath Manchester's leaden skies.

From the indie pop bands of the knife-plunging city to the grunge musicians of Seattle or Charles Dickens writing in the stormiest years of London history, some writers and composers, poets and painters are famously inspired by despair. It might come from outside—*la grisaille,* as Parisians call their stylishly overcast sky. Or it may churn from within, temporary blues or serious depression. Often, it seems to be a combination—rain, sending gray streaks down windows and gloomy thoughts through the mind.

History's best-known musical rains patter in Frédéric Chopin's Prelude op. 28, no. 15, known as the "Raindrop Prelude." Chopin is believed to have written the piece, his longest prelude, during his stay at a Majorca monastery in 1838. The French novelist Amandine Dupin (better known by her pen name, George Sand), Chopin's mistress, who accompanied him to Majorca, recounted the night she returned to the monastery in a catastrophic storm, to find Chopin weeping as he played one of his new preludes. He was hallucinating, Sand wrote in her *Histoire de Ma Vie.* He saw himself drowned in a lake. "His composition that evening was full of raindrops that resounded on the roof tiles," Sand wrote, "but were translated in his imagination and in his chant into tears falling on his heart from the sky."

From the northern Atlantic Ocean, prevailing winds called the westerlies carry warm, moist clouds over the west coast of the British Isles. (To clarify a confusing argot of scientists and mariners: When they speak of wind, they refer to the direction from which it blows. Of an ocean current, they cite the direction toward which it flows.) As the clouds rise over the craggy landscape, they release the heaviest rains over Ireland and western Scotland, England, and Wales. This weather pattern makes small rural villages along the west coast the rainiest inhabited places in the United Kingdom. The wettest of all, soaked with 140 inches a year, is a charming fell-walking hamlet, Seathwaite, about two hours' drive northwest of Manchester.

Like the Cascades to Seattle, England's green and gentle Pennines stretch to the north and east of Manchester, holding perpetual light rain clouds overhead. Also like Seattle, Manchester is England's notorious city of rain despite not breaking the top ten rainiest places in the nation. People just seem to want a rain capital, and they want their rain capital to drip, not pour.

In their short and cult-worshipped history, the Smiths put out four albums, all of which reached the U.K. top five. The music is known for capturing British character and British weather. The Manchester-born music journalist Sarah Champion, in her book *And God Created Manchester*, wrote that Marr's psychedelic Bo Diddley riff in the Smiths' "How Soon Is Now" evokes the wet streets of Manchester "as powerfully as Ry Cooder's blues twang does the desert terrain of Paris, Texas."

Once Champion pointed it out, I could hear the rain in Marr's guitar. I thought of other bands and classic rock songs like the Doors' "Riders on the Storm," and the Seattle native Jimi Hendrix's "Rainy Day, Dream Away," and wondered how the musicians so well captured rain in mood and sound. Champion's suggestion of a rain riff and Bo Diddley, the American R&B legend who helped blaze the road from blues to rock, sent me to the phone to call my friend Karl Meyer, a blues bass player and producer in Chicago who is a big fan of Bo's music. Alternately talking and singing to me over the phone, he explained Diddley's signature beat, an Afro-Cuban rhythm with a three-two clave: BUM-BUM-BUM . . . BUM-BUM. In "How Soon Is Now," Marr plays the Diddley beat, but it's only the backdrop for the rain to come. For that, he tweaks the clave with the reverb on his guitar, creating an echo that, just like a storm, conjures a big sound rather than single drops. In the more overt rain rock songs, Karl tells me, a rainstorm can be as easy as turning up the reverb and hitting a Fender amp with a fist.

The Doors used those sorts of sound effects for the storm in "Riders." But the song's well-known rain solo comes from Ray Manzarek's long downward run on a Fender Rhodes electric piano.

Grunge has its own set of atmospherics. To find their origins, you have to drive two hours beyond Seattle, west to Kurt Cobain's birthplace, the logging burg of Aberdeen, Washington. Singer-guitarist Cobain and bassist Krist Novoselic famously formed the breakout band Nirvana in their hometown in 1987. There was definitely something in the air. Aberdeen makes Seattle look sun-drenched. With upward of 130 inches of rain a year—more than three times Seattle's annual catch—it is one of the wettest inhabited cities in the continental United States.

Like 1960s Manchester, it is not the rain that makes Aberdeen gloomy. It is the rain mixing with the yellowish smoke of a pulp mill, rusting through an abandoned railroad track, and spattering up from the tires of a log truck rumbling by on Highway 101. Aberdeen was the only rainy place I traveled where the color seemed to be draining away rather than brightening, a result of industrialization and its collapse. (Today's Manchester, with handsome converted warehouses, glass skyscrapers, and a thriving Gay Village, has largely overcome its grim past.) At the turn of the twentieth century, Aberdeen was a busy sawmilling town. By the turn of the twenty-first, some of its last big mills were being shuttered. The area consistently suffers the highest unemployment rates in Washington.

The day I visited, Aberdeen's horizon was indistinguishable between the gray sky and the gray harbor, appropriately named Grays Harbor. (For the one-eyed sea captain Robert Gray. He discovered the harbor in 1792 during a fur-trading expedition, but apparently did not bother to disembark.) The gunmetal monochrome extended to the aging industrial plants along the harbor, the rusting ships at berth there, and the fog that clouded it all. The small downtown was desolate, with the exception of one treasure trove, the Sucher and Sons Star Wars Shop—every inch crammed with intergalactic toys and collectibles, with Cobain memorabilia thrown in among the Chewbaccas and Clone Troopers.

Some ascribe the rise of Cobain and the flannel-swathed, booted army of young grunge musicians to the rain, or the slump in the logging industry, or the jilting of a generation—the first in American history to hear that they would never have it as good as their parents. The stormy electric sound also emerged in Seattle's isolation from New York and Los Angeles. Grunge's trademark rain and flannel are anti-hero symbols, real and authentic: no L.A. sunshine here.

Cobain always said how much he hated his gloomy hometown. But it also inspired some of his richest songs, like the gentle "Something in the Way," which captures his misery in drips from the ceiling of Aberdeen's Young Street Bridge, which he said he slept under as a teenager. Today, the bridge's concrete underbelly is full of tributes spray-painted to Cobain since he committed suicide at age twenty-seven.

As I read them, it seemed to me that giving rain credit for grunge, or Cobain's sadness or his songwriting, would be just the sort of misunderstanding that so tortured him in his short life. In the context of a storm, rain is just one stage direction in an atmospheric tempest. Maybe it's the same with the creative mind. Rain may not be the sole cause of the anguish or the art. But no doubt, it can create a mood and inspire a melody.

In her Kenya memoir *Out of Africa,* the Danish author Isak Dinesen described a diversion she created one evening while harvesting maize with young Swahili laborers. As a way of amusing herself and them, she put together Swahili words in rhyming verse. The boys formed a ring around her and waited eagerly for the rhymes, laughing each time she came up with one. "I tried to make them . . . find the rhyme and finish the poem when I had begun it, but they could not, or would not, do that, and turned away their heads. As they had become used to the idea of poetry, they begged: Speak again. Speak like rain."

Perhaps more than in music or any other genre, rain, so fit for meter and metaphor, speaks in the language of poetry. Anthologies seem to have no end of poems titled "Rain," or those devoted to April rain, May rain, August rain, September rain, summer rain, noon rain, night rain, and London rain—and all of that not even counting showers.

Conrad Aiken's beautiful "Beloved, Let Us Once More Praise the Rain" is emblematic of poetry that reveres rain in spirit and deed. The Pulitzer Prize–winning poet called raindrops "the syllables of water." Few worked the syllables as well as rain-loving Henry Wadsworth Longfellow. In his "Rain in Summer," written in 1845, streets and lanes relish the rain after a hot day. Rain relieves a sick man. It brings joy to boys who "down the wet streets / Sail their mimic fleets." In the country, rain is welcomed by the thirsting grain, the "toilsome and patient oxen," and the grateful farmer. But only the poet can see rain in its full cycle, following raindrops to rainbows and graves alike. "From birth to death, from death to birth / From earth to heaven, from heaven to earth."

Longfellow also wrote what is arguably rain's most famous refrain, the closing lines of "The Rainy Day":

Into each life some rain must fall
Some days must be dark and dreary.

The popular use of the lines has left "The Rainy Day" known for the bleakest of messages, rather than the hope Longfellow intended when he wrote, "Be still, sad heart, and cease repining; / Behind the clouds is the sun still shining."

Emily Dickinson wallowed in the inevitability of Longfellow's closing stanza, judging by the number of times she quoted one line or the other in her correspondence. "It's a sorrowful morning Susie," she wrote her dear friend and future sister-in-law, Susan Gilbert, in 1852.

"The wind blows and it rains; 'into each life some rain must fall,' and I hardly know which falls fastest, the rain without, or within."

Dickinson could write brilliantly of rain on her own, but that didn't mean she loved it; quite the opposite. (The belle of Amherst could, after all, write brilliantly of anything, including the common fly, with its "blue, uncertain, stumbling buzz.") Comparing raindrops with pearls in her poem "Summer Shower," Dickinson imagined, "What necklaces could be!" But beyond the warm drizzles that quenched her garden in summer and tamped down the dust that swirled from the dirt roads of Amherst, Dickinson viewed storms, and especially thunder and lightning, with a deep sense of doom.

"The Storm" gives a glimpse of her foreboding:

There came a wind like a bugle
It quivered through the grass
And a green chill upon the heat
So ominous did pass
We barred the windows and the doors
As from an emerald ghost
The doom's electric moccasin
That very instant passed.

Dickinson considered rain a stark reminder of loneliness, and a spoiler of fun. In June 1851, when she was twenty, she wrote to her brother, Austin, on a Sunday evening just after he'd left Amherst to teach in Boston. She sat beside the family hearth as a storm blew from the northeast and her mother complained of icy feet. "We are a rather crestfallen company," Dickinson wrote, ". . . what with the sighing wind, the sobbing rain, and the whining of nature *generally*, we can hardly contain ourselves, and I only hope and trust that your this evening's lot is cast in far more cheery places than the ones you leave behind."

In other letters, Dickinson describes her horror at surprise rainfall during a carriage ride "in drops—sheets—cataracts—what *fancy conceive* of drippings and of drenchings," and, on another occasion, "shower after shower of chilly pelting rain" that kept her from leaving her family's home for church.

Later in life, she would not venture out on sunny days, either. Perhaps the rain was an early, convenient source of blame for her reclusive tendencies, which modern researchers have linked to panic disorder or agoraphobia. But gray, gloomy weather also appears to have served Dickinson as a creative force. She is the only writer for whom psychologists have evidence to bolster what is otherwise the conjecture of literary criticism.

The researchers believe that as a young woman, before her mental state became more acute, Dickinson suffered from Seasonal Affective Disorder. Known as SAD, the seasonal depression usually occurs in the gloom of winter, during months of scant sunlight. Studies have found that Dickinson cranked out many more poems in spring and summertime than in autumn and winter. But cognitive psychologists Christopher Ramey and Robert Weisberg asked a more relevant question. They set out to measure the quality of Dickinson's work in different seasons, rather than the quantity. Ramey and Weisberg calculated by season the portion of poems Dickinson wrote that were later collected in top literary digests and anthologies. During the years she was believed to have suffered from SAD, Ramey and Weisberg found that Dickinson wrote a considerably higher percentage of her *best* poems in autumn and winter, even though she wrote more poems in spring and summer. Dickinson's dark days, they concluded, "provided her with material that she could use in her poetry."

I n fiction, rain gives a sense of desolation and decay in virtually all the works of Charles Dickens, falling through a broken roof or "slowly and doggedly down" on Pickwick, "as if it had not even the spirit to pour." Rain is a sure mourner at a Dickens funeral. It is a protagonist in

Bleak House and a constant warning in *Hard Times,* seeming to caution the unhappily married Louisa against adultery when James Harthouse asks where they are to meet. Louisa was sure "there was another listener among the trees. It was only rain, beginning to fall fast, in heavy drops."

Dickens's imagination revved up in the dark, and he especially liked to conjure rain in the black of night. The late Dickens scholar Bernard Shilling wrote, "As the river in darkness is most heavily charged with suggestions of its mystery, its terrible implacability, its indifference and continuous energy, so does the rain achieve its meaning in the silence and loneliness of night." In *The Old Curiosity Shop,* when Little Nell's grandfather steals her savings, she rises from her bed in the dark night while "the rain beat fast and furiously without, and ran down in plashing streams from the thatched roof."

Climate scientists say it is no accident that Dickens is known for his desolate weather. He lived during the worst years of the Little Ice Age, the five hundred years of extreme cold and storms that lasted until about 1860. We can hardly imagine the conditions, though Dickens's work gives a most vivid picture. The Thames froze regularly and merchants set up "frost fair" carnivals on the ice. Snowy London winters were common then but no longer; Dickens's work shaped the cultural nostalgia for an old-fashioned white Christmas.

Another English classic for which we can thank the Little Ice Age is Mary Shelley's *Frankenstein.* She wrote it during a stretch of freezing rain while vacationing with her husband, Percy Shelley, and Lord Byron near Geneva in the summer of 1816. The eruption of Mount Tambora the year before dimmed the sun and brought what's known as "the year without a summer." It was the coldest summer ever recorded in Europe. Shelley and her poet companions had to stay holed up in their villa, huddled over a constantly burning fire. Lord Byron suggested they each write a ghost story. *Frankenstein* was hers.

For every writer who conjures rain "dismally against the panes" like Shelley or "cold, unending, heavy, and accursed," like Dante in his third circle of hell, there is another who finds beauty in rain's silver guise and wonder in its service to nature. Such was the outlook of the English novelist Walter Raymond, who is credited with the metaphor "right as rain." The phrase first appears in Raymond's *Love and Quiet Life*, published in 1894. Etymologists surmise it took hold for reasons of alliteration and not logic. The previous iterations: right as a ram's horn, in the fourteenth century; right as a line, in the fifteenth; right as my leg, in the seventeenth; and Dickens's right as a trivet simply weren't as catchy.

Raymond wrote in an old, thatched-roof cottage in a village called Withypool on the wet southwest coast of England, now preserved as part of Exmoor National Park. During breaks he would take long walks along the moorlands and the river Barle. He wrote exquisitely of rain in his *Book of Simple Delights*. The Earth, and he, were grateful when rain would fill "the whole atmosphere with a healthy freshness, that the mingling scent of all the flowers can never cover or excel."

I like to think the etymologists are wrong—that the saying "right as rain" made perfect sense to Walter Raymond, and makes perfect sense today. Count him among the artists who simply loved rain. He thought it was just right.

As a plot device, rain is good at forcing people together, sometimes cozily, sometimes uncomfortably, a well-known impresario in Thomas Hardy's 1883 short story "The Three Strangers." Three mysterious men, one after the other, ask to join a party at a shepherd's cottage to take shelter from a rainstorm that "smote walls, slopes, and hedges" and blew little birds' tails "inside-out like umbrellas." They turn out to be an escaped condemned man, a hangman, and the escapee's brother. After the strangers manage to give everyone at the party the creeps and two of them leave, the hangman and a constable figure out they've let an escapee slip by and the party gives chase, only to seize the wrong man—the brother. No one hangs in the morning. And rain has helped the story hold up for more than a century—sealing the plot, lending a mysterious atmosphere, and symbolically, for the deeply religious Hardy, putting hangman and condemned man on equal footing: Rain falls upon the just and the unjust.

Often, rain's appearance on page or screen signals misfortune or wickedness to come. Ernest Hemingway's 1929 novel *A Farewell to Arms* is full of rain as metaphor for foreboding and doom, fate and death—but also longing. After Hemingway kills off his protagonist Frederic Henry's lover in childbirth, he completes the circle of calamity with this final sentence: "After a while I went out and left the hospital and walked back to the hotel in the rain."

Hemingway wrote forty-seven alternate last lines for the book before he settled on Henry's rainy walk back to the hotel. Several were too sunny, many involved rain. None worked as perfectly as the one he chose—rain standing in for the tears of a man unable to express his mourning.

Rain is also good at cleansing characters of their faults. The American writer Edward Lewis Wallant, compared with Bellow and Roth before he died in his thirties, does it in his novel *The Pawnbroker,* foreshadowing a troubled young character's redemption with a walk in a storm: "The fiery exultation of evil drained out of him then, and he walked home, all hunched over, nailed heavily to the earth by the torrential downpour."

The counterpoint to rain as cleanser: It can just as easily coat you in mud. In *Song of Solomon,* Toni Morrison has her poor jilted character Hagar spend her mother and grandmother's last two hundred dollars on a shopping spree to win her lover Milkman back. To try to transform herself into the sort of woman (a lighter/whiter ideal) he wants, she splurges on bra and panties, garter belt, nylon slips, heels, a suit, a blouse, Youth Blend crème, jungle-red lipstick. A manic walk home in a rainstorm rips her hose, soils her white suit, lumps her face powder, streaks her rouge, and sends her hair into "wet, wild shoals." She dies of a fever not long after. The sight of herself sullied and bedraggled fills her eyes "with water warmer and much older than the rain."

The Seattle-based writer Timothy Egan, who wrote a lyrical travelogue of the Pacific Northwest called *The Good Rain*, once did his own, informal study of book authors in his city to figure out if they accomplished their best work in the murk of the dark months. "Creativity needs a season of despair," he wrote. "At the calendar's gloaming, while the landscape is inert, and all is dark, sluggish, bleak and cold, writers and cooks and artists and tinkerers of all sorts are at their most productive."

As he tracked down Seattle's writers behind their laptops and in the latte shops, Egan's theory panned out. The authors he interviewed felt "an overpowering impulse to write" in the wintry days of gloom and scant sunlight. Seattle transplant Jennie Shortridge said she took seven years to finish her first novel in Denver, with three hundred days of annual sunshine. "When I moved to the Northwest, I wrote the next novel in fifteen months, and subsequent books every two years," she told Egan. "The dark and chill keeps me at my desk."

Literary inspiration turns on the type of rain, says Thomas Hallock, a professor of literature in Florida, where the sunshine ratio makes it notoriously hard for writers to stay at their desks. While Seattle and Manchester set a misty mood, "big storms are a whole different story," Hallock told me, the sort to inspire the wreckage in Zora Neale Hurston's novel *Their Eyes Were Watching God*. The title was inspired by a 1928 hurricane that sent South Florida's Lake Okeechobee bursting through and over its earthen dike, killing 2,500 people, most of them poor black laborers who drowned in the agricultural fields south of the lake.

Hurston had her storm outrage and Hemingway his; he wrote his essay "Who Murdered the Vets?" in response to the Great Labor Day Hurricane of 1935 that ravaged the Florida Keys and killed some five hundred WPA workers, half of them World War I veterans sent to build a highway across the low-lying islands during hurricane season.

If tropical weather inspires dramatic bursts at the keyboard, steadier drizzles explain Egan's consistently creative Seattle—and perhaps why Iceland produces more authors and books per capita than anywhere else in the world. It is said that one in ten Icelanders will publish a book. Clouds cover the capital city, Reykjavik, 90 percent of the time. The probability of rain (usually moderate showers or drizzles) rarely falls below 70 percent. The cozy national pastime of writing and reading (you can buy books at the gas stations) dates to the thirteenth and fourteenth centuries, when unknown authors put the famous Icelandic sagas to paper during an otherwise dreary, broke, and isolated time in the country's history.

What the Irish call soft days helped build the canon on their well-watered island, where contributions to literature far eclipse the nation's relative size. Rain and its mordant shades gave James Joyce a gripping suspense. In *Dubliners,* his boy narrator hears rain "impinge upon the earth, the fine incessant needles of water playing in the sodden beds." In his four-part "Gifts of Rain," it gave Seamus Heaney layers of antediluvian clay to search for a lost home, lost past, lost language, and lost way of life. The poem of water wordplay was a favorite of his to read aloud and a favorite of critics, who called it some of his best civil war writing.

Rain, gray skies, and lightning are constants in the work of Samuel Beckett, who loved to let just a tiny gleam of light into his rainy landscapes, his literary biographers note.

In his film *Radio Days,* a tribute to the glory days of radio and the Rockaways in the early 1940s, Woody Allen shows off his old beachside neighborhood on the dreariest day possible, a day of rain-sullen sky and sea, icy wind, and furious waves pummeling shore. "Forgive me if I tend to romanticize the past," Allen narrates. "I mean, it wasn't always as stormy and rain-swept as this. But I remember it that way . . . because that was it at its most beautiful."

The audience always laughs. And this Allen doesn't understand, he says in his book *Woody Allen on Woody Allen*: "I was serious. To me, it's beautiful. I'm always filming exteriors when it's dreary out. If you look at all my films over the years, you'll find it's never sunny, it's always gray. You would think that it rained in New York like London. That it's always grey and bleak in New York. I love the idea of rain. I just think it's so beautiful."

In film, rain adds stylistic pizzazz to its literary metaphor: tears, soul-cleanser, foreshadower of doom, equalizer. The last was a favorite of the Japanese director Akira Kurosawa, who seemed to conjure rain in every movie, including *Seven Samurai*'s epic closing battle. The Italian-born American director Frank Capra loved rain for the sensual power it could bring his films, including *It Happened One Night* and *It's a Wonderful Life*. Rain, Capra once said, "is for me an aphrodisiac."

When Allen has written rain into a scene and good weather foils a filming day, he covers the set with rainmaking machines. Movie rainmakers look like the hulking linear crop irrigators that crawl along thirsty fields of soybeans in the Midwest—but hoisted high into the air on a crane. For lower-budget filmmakers, special effects companies also rent smaller rain towers; rain wands that can be hand-held over actors for close-ups; rain bars that can attach to a car roof; dump tanks that can pour 850 gallons of water on an unlucky actor's head; or a rain window that, when attached to a hose, creates the lonely specter of rain streaming down a glass pane.

Filming with rain machines is an order of magnitude more complicated than other outdoor scenes. The machines add thousands of dollars a day to production. Their generators grind loudly and can drown out the actors, adding considerable editing work on the back end. And for every take, actors have to blow-dry their hair, don a new set of clothes, and steel themselves for the next deluge. The final, rain-kiss scene in *Breakfast at Tiffany's* required eight takes and two dressing rooms for star Audrey Hepburn, "Wet Hepburn" and "Dry Hepburn."

Still, Allen says he regrets the times he's let producers talk him out of the aggravation. When it rains: "People are confined to their households. They seek shelter. They succor inside their houses. They

run from the outside to the inside to protect themselves. They go inward and move inward."

As certain as it is to rain on any human pursuit eventually is the
certainty of romance in the rain, at least in the movies. Tap dancing
in puddles on a city sidewalk in the most recognizable rain scene of
all time, a buoyant and giddy Gene Kelly mocks the claustrophobic
storms of a Woody Allen film in the 1952 musical *Singin' in the Rain.*

Singin' *to* the rain as much as in it, stretching his arms to embrace
it, removing his hat to lift his face to it, Kelly seems equally in love
with rain as with his ingénue, played by Debbie Reynolds. The film
celebrates rain as playful and uplifting, its singularity and suddenness
much more romantic than the sun.

A young Irish filmmaker named Claire Dix appreciates the rain
as joyfully in her short ode to Ireland and rain, *Downpour.* The three-
minute film opens with a bride putting on her wedding dress as rain
streams down the windows outside. While her mother frets about the
rain on the telephone in the background, the bride is happy and serene
remembering all the rainy times that have defined her romance with
the man she is about to marry: their meeting at a bus stop during a
deluge, swimming in the rain at the beach, camping in the rain, and
making love with the rain drumming outside.

Dix grew up in West Cork on Ireland's southwest coast, where gale-
force winds and thundery downpours are as much part of the landscape
as the jagged green coastline. She hit upon the idea for *Downpour*
while brainstorming a funding project offered by the Irish Film Board.
Entrants had to come up with a short film with the theme "Ireland,
I Love You." "Rain makes Ireland the country it is, and *Downpour* is
meant to celebrate our love-hate relationship with rain," Dix told me.
"I thought to myself, 'What day would be universally recognized as
being ruined by rain?' The wedding day clicked."

Downpour racked up numerous awards, from the Irish Playwrights
and Screenwriters Guild prize for best short script to the Directors
Choice Short Film Award at the Irish Film Festival in Boston. The

accolades floored Dix, who felt a bit sheepish setting a romance in the rain. She knew she risked coming across like a giant Irish cliché.

Rain is as romantic as roses, and as easily overdone, like in the famously torrential scene at the end of *Four Weddings and a Funeral*. As rain pelts her eyes, a sopping Andie MacDowell tells Hugh Grant: "Is it still raining? I hadn't noticed." The unconvincing line is regularly voted one of the worst movie lines of all time. The kiss that follows, however, is among the popular rain kisses of film history, up there with Kelly and Reynolds's in *Singin' in the Rain*; Hepburn and George Peppard's at the end of *Breakfast at Tiffany's*; and many modern twists on the soaked smooch, upside down between Kirsten Dunst and Tobey Maguire in *Spider-Man* or preceding disaster with John Hannah and Gwyneth Paltrow in *Sliding Doors*. (Naturally following all that making out is the rain sex scene, never as memorable as Mickey Rourke and Kim Basinger's in the vaporous *9½ Weeks*.)

When Dix herself married in 2010, she and her fiancé took a chance on an outdoor wedding. They held it on Ireland's Sheepshead peninsula overlooking the sea. It was an unusually sunny day in West Cork. "It was amazingly sunny. The type of day we never have here," she says. "You couldn't write a screenplay in weather like that."

In the human relationship with rain, art's content reflects the duality of love and hate, cleansing and muddying, blessing and curse. The balance holds true for form, too. Rain can be avant-garde in a Beckett play and embarrassingly melodramatic in a romance novel—or when the rain machine gushes a bit too obviously in film.

In 1713, the satirist Alexander Pope poked fun at literary atmospherics with a sardonic recipe book for aspiring poets hungering to write an epic. To conjure a tempest, he advised: "take Eurus, Zephr, Auster, and Boreas, and cast them together in one verse. Add to those of rain, lightning, and of thunder (the loudest you can)." Finally: "Mix your clouds and billows well together until they foam, and thicken your description here and there with a quicksand. Brew your tempest well in your head before you set it a blowing."

The Victorian writer Edward Bulwer-Lytton didn't pick up on the irony. He composed several of the most enduring phrases in western culture, including "The pen is mightier than the sword" and "the almighty dollar." But the seven words he chose to open his 1830 novel *Paul Clifford*—"It was a dark and stormy night"—earned him literary immortality for bad writing. Charles Schulz's Snoopy may have done the most to solidify Bulwer-Lytton's reputation, tapping out the line over and over again on his dog-house typewriter.

The incipit is only part of the sentence, which goes on to read, "the rain fell in torrents—except at occasional intervals, when it was checked by a violent gust of wind which swept up the streets (for it is in London that our scene lies), rattling along the housetops, and fiercely agitating the scanty flame of the lamps that struggled against the darkness."

American humorist Mark Twain opens his 1894 novel *The American Claimant* by announcing to his readers, "No weather will be found in this book. . . . Many a reader who wanted to read a tale through was not able to do it because of delays on account of the weather."

Twain refers readers to an absurd appendix where they can pick and choose from a list of purple weather descriptions: "Wild piles of dark and coppery clouds, in which a fierce and rayless glow was laboring, gigantically overhung the grotesque and huddled vista of dwarf houses, while in the distance, sheeting high over the low, misty confusion of gables and chimneys, spread a pall of dead, leprous blue, suffused with blotches of dull, glistening yellow, and with black plague-spots of vapor floating and faint lightnings crinkling on its surface."

Twain had prophesied a British pop star's tongue-in-cheek miserabilism a century before he thought of it himself.

A music writer once asked young Steven Patrick Morrissey: If forced to leave England at gunpoint, where would he go? Morrissey answered Guernsey or Jersey, two bailiwicks that make up the Channel Islands. In the possession of the British Crown, they are about as close as you can get to the U.K. without technically being there.

"Not Los Angeles?" the reporter asked.

"No. I need grit and struggle and Los Angeles is terribly nice, but people, once they get there, cease to be real. Constant and repetitive fulfillment is not good for the human spirit. *We all need rain and good old depression.* Life can't be all beer and skittles."

To the dismay of their loyal fans around the world, the Smiths broke up in 1987, after only five years of making music. Morrissey and Marr never spoke again. Morrissey, tapping his trademark melancholy, soared with his first solo album, *Viva Hate.* He was dubbed the Pope of Mope. On his own, he made the top ten of the U.K. singles chart on ten occasions. Yet his subsequent albums have sold fewer and fewer copies.

I could not help but wonder if Morrissey's music was missing dark and drizzly Manchester. As the millennium turned, he had moved to a new estate, in a new part of the world. Newlyweds Clark Gable and Carole Lombard purchased the bucolic place in 1939 and fixed it up together, but they enjoyed it for only three years before Lombard was killed in a plane crash at the age of thirty-three. F. Scott Fitzgerald was in the neighborhood, too: He rented a little guesthouse near the estate, in a heartbreaking and unproductive stretch of screenwriting that preceded his death.

The estate lies just south of Ventura Boulevard, in sunny, rain-spare Los Angeles.

THE SCENT OF RAIN

n the human quest to capture rain, the artists and writers trump the engineers, their images and words growing stronger while dams weaken over time. But the most wondrous attempt to catch ephemeral rain is made by villagers in an outpost of northern India. It involves not a book or a barrier, but a tiny bottle.

In India's state of Uttar Pradesh, the village of Kannauj lies a dusty, four-hour drive east of the tourist-mobbed Taj Mahal, white-marbled wonder built by the Mughal emperor Shah Jahan in memory of his third and favorite wife. Empress Mumtaz Mahal died in 1632, giving birth to their thirteenth child.

The Taj is Jahan's grand paean to lost love. But he also mourned his queen in much more personal ways. For a time, he traded his colored and embroidered garments for bereavement white, and he abandoned music, a sacrifice for a devoted patron. While color and song eventually returned, Jahan never again wore perfume. Fragrant oils—known in India as *attars*—had been one of the couple's great shared passions.

Long offered to the gods, aromatic oils and incense during empire days became the purview of royalty, too. A Portuguese friar named Sebastien Manrique, smuggled by a eunuch into a dinner of elites one evening to spy on Shah Jahan, wrote with astonishment of the gold vessels, silver braziers, and perfume holders emanating the scents of ambergris, eagle wood, and civet. A seven-spouted silver hydra spewed scented water into a trough.

Then and now, Kannauj was the place to fetch the fine scents—jasmine oils, rose waters, the roots of grasses called vetiver, with a bouquet cooling to the nose. Exactly when attar-making began here, no one is certain. Archaeologists have unearthed clay distillation pots from the ancient Harappan ruins of the Indus Valley. Fragrance references also are found in the Hindu holy text the Rig Veda.

In the seventh century, a king called Harshawardhan made Kannauj the throne of his north Indian kingdom. The industry must have been well under way then, because he was taxing vetiver. But after his death, his empire disintegrated into small, rival states. In the millennium and a half since, Kannauj's royal past has been largely forgotten. But its attar industry has quietly thrived, growing to become the largest in India.

Today, Kannauj is to India what Grasse is to France, hub of a historic perfumery that draws much of the town to the same, aromatic pursuit. Most of Kannauj's villagers are connected to fragrance in one way or another, from sinewy craftsmen who steam petals over wood fires in hulking copper pots, to mothers who roll incense sticks in the shade while their toddlers nap on colorful mats nearby.

But India's fragrance capital has something Grasse and the other aroma-cultivating places of the world do not. Along with their ancient perfumery, the villagers of Kannauj have inherited a remarkable skill.

They can capture the scent of rain.

S mell," wrote Helen Keller, "is a potent wizard that transports us across a thousand miles and all the years we have lived." In her perceptive essays that helped prove the deaf and blind could master language and much more, Keller described scent as the fallen angel of the senses. People appreciate it enough in a lovely garden, but too often ignore its complexities.

With her nose alone, Keller said, she could tell a man from a woman and distinguish among professions—carpenter from ironworker, artist from mason. Most powerfully, scent preserved her strongest memories. "A whiff of the universe makes us dream of worlds we have never seen, recalls in a flash entire epochs of our dearest experience. I never smell daisies without living over again the ecstatic mornings that my teacher and I spent wandering in the fields, while I learned new words and the names of things."

And from the time of her childhood, Keller would never forget the smell of a coming storm.

> *My little friends and I are playing in the haymow. The sense of smell has told me of a coming storm hours before there was any sign of it visible. I notice first a throb of expectancy, a slight quiver, a concentration in my nostrils. As the storm draws nearer, my nostrils dilate the better to receive the flood of earth-odors which seem to multiply and extend, until the splash of rain against my cheek.*

Every storm blows in on a scent, or leaves one behind. The metallic zing that can fill the air before a summer thunderstorm is ozone (from the Greek verb *ozein,* "to smell"), a molecule formed when electrical discharges, in this case from lightning, break oxygen's two atoms down to three. Likewise, the familiar, musty odor that rises from streets and storm ponds during an old-fashioned deluge is called geosmin. A by-product of bacteria, geosmin also gives beets their earthy flavor. But what's pleasant in rain and root veggies can ruin a cool glass of water or a catfish fillet. Geosmin is the bane of urban water suppliers and freshwater fish farms. The two products don't go over with too much *gout du terroir,* taste of the land.

Rain also picks up odors from the molecules it meets. So its essence can come off as differently as all the flowers on all the continents—rose-obvious, barely there like a carnation, fleeting as a whiff of orange blossom as your car speeds past the grove. It depends on the type of the storm, the part of the world where it falls, and the subjective memory of the nose behind the sniff.

City rain smells of steaming asphalt, in contrast—not always unpleasantly so—to the grassy sweetness of rain in the countryside. Ocean rain smells briny like Maine clam flats on a falling tide. In the desert Southwest, rare storms punch the atmosphere with creosote and sage. In the Southeast, frequent squalls leave the damp freshness of a wet pine forest. "Clean but funky," Thomas Wolfe called the exquisite scent of the American South.

But nowhere is rain's redolence more powerful than at the climatic extremes of the world, where great, dry swaths of desert are inundated with the most dramatic seasonal storms on Earth. The monsoons of India, Southeast Asia, western Africa, and parts of Australia can turn deserts to grasslands and famine to fortune. In the otherwise dry places that depend on the downpours for most of their annual rainfall, the monsoons shape everything from childhood to culture to commerce. And they arrive with a memory-searing scent.

To Sanjiv Chopra, the Indian American Harvard Medical School physician and author, like his younger brother Deepak Chopra, the loamy smell of long-awaited rains soaking India's thirsty ground is "the scent of life itself."

The British psychoanalyst Wilfred Bion, born in northwest India, credited the monsoons of his childhood with having given him a lifelong sense of rain as "the great event." He left India when he was eight. At fifty-three, he wrote of how his mind could still "recapture the thrill of the smell of parched land rain-soaked."

The earthy essence is strongest when rain quenches dehydrated ground. The scent can so tantalize drought-stricken animals that it sets thirsting cattle walking in circles. Not infrequently, writers and poets mine the language of fertility or sexuality to convey the aroma. A leading Aussie poet, Les Murray, describes a first rain's pheromones as the "sexual scent of Time itself, philter of all native beings."

In the 1950s and '60s, a pair of Australian mineralogists, Isabel Joy

Bear and Richard Grenfell Thomas, set out to discover the source of that piquant perfume. They dug it up from the earth itself, baked by the sun into rocks and clay. Using steam distillation, they managed to extract the scent, a three-part chemical compound that clumped into oily, yellow globules. One simple, one acidic, and one neutral with fatty acids and other organics, no single of the three could produce the scent alone. Just like perfume, the key to the redolence was in the range.

Ultimately, Bear and Thomas linked the scent to organic compounds that build up in the atmosphere, including heady-smelling terpenes secreted by plants. The major components in turpentine and resin, terpenes also put the essence in essential oils. They are the freshness in pine, the cool in peppermint, the spice in ginger. From the tallest conifers and from the tiniest mosses, hundreds of millions of tons are released into the atmosphere each year. Unleashed, the terpenes make some remarkably diverse contributions. They give hops its bite and cannabis its smooth character. They help form the blue haze that hangs over the Blue Ridge Mountains of Virginia and the Blue Mountains of Australia. They also make some of the planet's most intoxicating perfume.

Rocks and clay absorb terpenes and other molecules from the atmosphere like sponges: the drier they are, the more they draw in. Meanwhile, the hotter the temperatures, the faster the ingredients cook into Bear and Thomas's integral odor. During hot, dry stretches, desertlike places build up great stores of the compound. When the humidity shifts ahead of the monsoons, moisture loosens the material from its rocky pores and sends its pungency adrift on the wind. This is why sometimes rain's bouquet blooms before a storm, and sometimes after. The aroma is more powerful in the wake of drought because the essential oils have had longer to build in the layers of rock.

Publishing in the journal *Nature* in 1964, Bear and Thomas proposed a name for the scent brought on by rain. They called it "petrichor," a blend of the Greek words *petra,* rock, and *ikhor,* the ethereal fluid that flowed as blood in the veins of the gods.

But the scientists acknowledged that they were not the first to identify the stormy smell. They were not even the first to extract it. In fact, the element they dubbed petrichor was already the signature fragrance in an attar produced in an ancient perfumery found in the north Indian state of Uttar Pradesh, in the village of Kannauj.

The scientists flubbed the name of the fragrance, which is called *mitti ka attar.* But they got the translation right: Extracted from parched clay on the eve of the monsoons, and distilled with techniques dating to the Harappan, the scent of rain, in India, is known as Earth's perfume.

ndia's cleansing monsoons may stir a beloved national aroma, but before they arrive, the capital, Delhi, often reeks of some of the fouler ones—rotting garbage and excrement, especially at the Old Delhi Railway Station, where I wait on a hot June evening to board an overnight train to Kannauj. Each time a train pulls out, small, barefoot boys jump onto the tracks to pick through the swill left behind, among the estimated half-million ragpickers in Delhi who handle the vital work of garbage collection with no official pay or protection.

The children seem too focused on the task to be bothered by the waste and stench from the trains' open-pit toilets. With those odors sending my train into the night, the distinctness of the village of Kannauj will lie, indelibly and deliciously, in its scents.

On the outskirts of the city, truck-farm-sized fields planted with aromatic crops stretch for miles, interspersed with the monolithic chimneys of hundreds of small-scale brick kilns for which the region is also known. Like the attars, bricks are manufactured here little differently today from centuries ago, red clay earth cut from topsoil, then stacked and fired by men whose fathers, grandfathers, and great-grandfathers cut, stacked, and fired bricks too.

In the crop rows, white jasmine flowers shaped like starfish bloom in their ocean of waxy dark green. Twiggy trees called *gul-hina* are blooming too, their tiny flowers clustered into points of white flame. Ordinary on the tree, *gul-hina* leaves become the extraordinary henna

that decorates women's hands and feet for special occasions, or tints
dark hair a spicy red. Less well known: Its flowers make a delicate-
sweet attar.

It can take about one hundred pounds of flower petals or herbs,
infused into a pound of sandalwood oil—the ideal and purest base for
essential oils—to make one pound of pure attar. Extended families
head out in the early mornings or cooler evenings to pick the delicate
flowers. They pack their harvest in jute sacks, then rush, before the
petals start to wilt, to one of two dozen steam distilleries in the town.

But I've come to Kannauj for the only attar made from neither flowers nor plants, one manufactured nowhere else in the world. Never
have I traveled so far on such little assurance that I'll find what I'm
looking for.

I t was springtime when I began to correspond with Shakti Vinay
 Shukla, director of India's Fragrance & Flavour Development Centre in Kannauj. He confirmed that locals still bottle the scent of rain,
just as described in Bear and Thomas's half-century-old paper. But his
e-mails were guarded, reflecting the secrecy that surrounds perfumers
and the embattled psyche of the natural attar industry, which is losing
a long and painful war with synthetics.

After flying 8,000 miles to India, taking a train to rural north-
central Uttar Pradesh, catching an auto rickshaw to the fragrance-
center boondocks outside Kannauj, and finally arriving in the morning
at Shukla's office on the funky agricultural compound, I still had only
three worrisomely vague answers: Yes, *mitti attar* is still manufactured
in Kannauj, and only in the heat of May and June, when petrichor has
built up but the monsoons have not yet begun. Yes, Shukla would show
me the *mitti*-making, but "only once we are sure about your identity
and mission." And yes, there would be a place for me to sleep in Kannauj, at the fragrance center's government-run hostel, as long as said
identity and mission checked out.

When I arrive, the 26-acre campus is flourishing with medicinal and aromatic plants (exactly fifty types of each), even in the premonsoon melt. I pass human-sized clumps of lemongrass, citronella, and the fragrance staple vetiver, whose cooling roots are woven into curtains and hats in India to fend off summer heat. Under a sprawling open-air barn, menthol mint is drying in tall piles next to a portable field furnace that will distill 1,200 kilos of herb into 12 kilos of oil (grossing $12,000 on $1,000 in production costs).

In a complex of laboratories, scientists and grad students are researching plant remedies for everything from memory loss to malaria. At the Soviet-looking hostel softened by white-blooming frangipani trees, farmers from across India bunk for weeklong stints to train in aromatic plant cultivation.

As it turns out, Shukla needs only to look me in the eyes to decide I am no industrial spy. He asks his joyful round chef, Babu, to prepare my room in the hostel and some breakfast, potato *parotta* with yogurt and spicy pickled lime. Afterward, we set off in his 1950s-looking car, a black Hindustan Ambassador Classic, to chase the scent of rain on a cloudless day.

When the British novelist Neil Gaiman got to write an episode of the BBC sci-fi phenomenon *Doctor Who*, he made "petrichor" a telepathic password for the Doctor's time machine. On the trail of the scent in old Kannauj, the word also transported me—to an ancient city that has never let go of its past.

In modern times, Kannauj is also the name of a political district, sprawling home to more than 1.5 million people in Uttar Pradesh. But the old city retains much of its aromatic history—an estimated 40,000 of its 70,000 residents are engaged in fragrance in one way or another—and all the eccentric authenticity typical of Indian villages.

Small houses and perfume shops, made of a hodgepodge of materials, are packed side-by-side on the lanes. A brick house with a thatched straw roof butts up against a stone storefront with a corrugated metal roof. Colorful one-person Hindu temples are tucked here and there to honor gods. Cows wander the road, oblivious to their sacred standing. Bicycles loaded perilously high with bundles of incense sticks wobble by. A donkey pulls a cartload of flower sacks rather than flour sacks. All seem to ignore the tiny taxis and oversized work trucks decorated with Hindu icons and honking urgently to pass.

Stretched across the Makrand-to-Kannauj road, a brick archway carved and painted with colorful flowers and vines is now covered with real ones. Erected in 1944, the temple-shaped edifice announces the business of Kannauj in Hindi and Urdu: "Perfumes, Scented Tobaccos and Rose Waters."

Villagers say the Makrand-to-Kannauj road once smelled like heaven—covered on both sides with the shavings of sandalwood, a tree as central to Indian religion and culture as it is to the fragrance industry. That was before the near-decimation of India's sandalwood forests. Shukla's scientists have various experimental crops planted at the institute's grounds in search of a sustainable forestry solution for the slow-growing prize.

The scientists have told me that Shukla is a supersmeller ("the nose of all noses," said one), trained in the European perfume industry and committed to Kannauj more as missionary than bureaucrat. He is pained to watch his native country's attar industry lose market share to modernity. When India opened its economy to foreign trade in the early 1990s, brand-conscious young Indians began turning to French perfumes. For the past decade or so, the industry has survived on attar's popularity as a fragrance for tobacco products and a chew, *pan masala*. But with many states calling for bans on the cancer-causing products, reliance on this single market may not be possible.

Shukla seems to know the names of everyone in the village, along with those of their children. His wife, an obstetrician at the public hospital here, has delivered many of their babies. Beyond the archway and down a dirt road where the houses, bicycles, livestock, and all else are covered with a chalky layer of dust, we arrive at the home of a family called the Siyarams. For generations, they've made their living from an earthen pit behind their home. Covered with rainwater in the monsoons, the pit has dried out in the premonsoon summer; they bring water as they need it from an adjacent pond.

The Siyarams used to be known for making disposable clay cups popular in India called *kulhad*. Street vendors serve tea, *kulhad wali chai*, in the wee unglazed cups. Indians drink the tea and toss the *kulhad*, which break street-side with a brittle clunk. Tea sipped from the special cups is infused with an earth-rich taste and smell. The scent is the same sought by perfumers for *mitti attar*, and comes from the same source—clay flooded with rainwater during monsoon season that parches into a fragrant chalk in the dry times.

The perfumers of Kannauj once recycled *kulhad* pieces to make *mitti attar*, but demand for the broken cups outstripped supply. About twenty-five years ago, the Siyarams realized they could move up the supply chain, selling their scented earth to perfumers by the wagon-load rather than hawking individual cups from door to door.

Like the other fragrances manufactured in Kannauj, *mitti attar* begins with a harvest. In the Siyarams' clay pit, Mom, Dad, and their grown children are squatting in ankle-deep slurry, using their hands to shape parched yellow marl into clay disks. Like so much of rural India, the scene is a blend of traditional and modern, all against a premonsoon backdrop of heat and dust. The patriarch wears a skirt called a dhoti, his wife, a flowery sari and head scarf; their sons and daughters wear pants. They use wooden sticks to break the parched earth, and a diesel pump to draw water from the pond to wet it for clay. After they make the disks, they pile them for firing in a primitive kiln, dug into the side of the pit and covered with bricks and straw. Atop the ancient-looking pile, a rainbow of mobile phones gleams in the sun.

Baked and ready for Kannauj's perfumeries, the clay disks are called *khapra*. The next time I spot them, they are heaped in a dark corner of a cavernous fragrance distillery. Back through the old city of

Kannauj, we've followed narrow, winding roads to a perfumer called Munna Lal Sons & Co. I meet the third-generation leader of the company, Akhilesh Pathak, and a member of the fourth generation—his daughter, Swapnil, a twenty-four-year-old engineering graduate who grew up at boarding school and has just returned to Kannauj to learn the family's fragrant trade.

Each generation has built part of the eclectic complex where the extended family also lives in a row of well-appointed white houses. A content-looking herd of water buffalo lounges in the shade of a pair of massive Indian lilac trees that separate the homes from the perfume-making. Pathak says his grandfather Munna Lal made the rain fragrance from the time he opened for business in 1911; Lal taught the techniques to Pathak's father, who taught them to him.

Lal built the original two-story perfumery, now a faded yellow and cyan and crawling with shy monkeys peeking down from the balconies. In 1962, Pathak's father built the brick distillery where the company brews essential oils including *mitti attar*.

If Kannauj feels last-century, stepping into the distillery is more last-millennium. There is no artificial lighting, no machinery, no trace of modernity. It's like a medieval fort, with dirt floors and columns supporting the partial roof. Through the roof and open sides, natural light streams onto a primitive scene: Ropy craftsmen wearing only threadbare dhotis are tending fires under hulking copper cauldrons called *degs*, which, topped with oval lids, poke up from long rows of brick stills like giant fossilized eggs.

The ancient, painstakingly slow distillation practiced in Kannauj is called *deg-bhapka*. Each still consists of the copper *deg*—built atop its own oven and beside its own trough of water—and a bulbous condenser called a *bhapka* (receiver) that looks like a giant butternut squash. When a fresh supply of flowers comes in, the craftsmen pack a hundred pounds or so of rose or jasmine or other petals into each *deg*, then cover them with water, hammer the lid down on top, and seal it with a long rope of mud. They light a wood or cow-dung fire underneath. They fill the *bhapka* with sandalwood oil, and sink it into the trough. The *deg* and *bhapka* are connected with a hollow bamboo pipe called a *chonga*, which carries the fragrant vapors from the simmering pot into their sandalwood oil base.

Like the Siyaram and Pathak families, the distillery workers have inherited a precise skill from fathers and grandfathers. They must closely monitor the fires so the heat under the *deg* stays warm enough to evaporate the water inside to steam—but never so hot that it destroys the aroma. As fragrant steam travels from the copper pot through the bamboo pipe and into the squash-shaped receiver, the workers must keep the trough of water on the receiving side cool enough for the vapors to condense back into a liquid, infusing the sandalwood oil with their heady scent. Just by reaching their hands under the water to feel the bulbous part of the receiver, the perfume-makers can tell how much vapor has condensed. Every couple hours, they switch out the receiver, cooling down the *deg* with wet cloths each time to stop the condensation. A typical hundred-pound batch of petals takes six or seven hours to distill.

On this day, the distillers are brewing the only attar that doesn't come from a plant. Like feeding coal into a furnace, they shovel the clay disks into the copper pots before pouring in the water, hammering on the tops, and sealing them with the mud. Just like with the flower petals, the men will mind the *mitti* for six to seven hours before all of the aroma steams out of the clay. At that point, they'll drain the receivers from a hole in the bottom, siphoning off the water that has condensed in the vessel until only the rich, fragrant oil pooled on top remains.

The *mitti attar* is not finished until a final step, when it is poured into a special leather bottle called a *kuppi* and sealed inside. Attar not stored in the *kuppi* "is essentially ruined," says Shukla, ever wary of modern manufacturing techniques, especially anything to do with plastic. "The moment you put it in the leather bottle is important, like the moment you put it on your skin. It allows the attar to release any remaining moisture and realize its true scent, in this case, the first rain on the ground."

Back on the labyrinthine streets of old Kannauj, we find the shop of the man who makes the bottles, Mohammed Mustakin, who traces his craft lineage even further back than the Siyarams or the Pathaks. Like in those family businesses, the next generations are here, too, working beside Mustakin, whose long white dress, or *kurta,* matches his squared white beard. I meet Mustakin's two sons, and a beautiful toddler grandson peeking from behind his dad, who translates for Mustakin: "My father and forefather and forefather and all the forefathers we remember made leather vessels for the attar," he tells me. "We always learned that our vessels are the same from the fairy tales of Ali Baba."*

Our last stop on the trail of the *mitti attar* is another Kannauj storefront, a retail perfumery owned by a three-thumbed shopkeeper named Raju Mehrotra. Also carrying on the business of his father and grandfather, Mehrotra sits at a soapstone counter, metal shelves behind him jammed with glass bottles and tins of every size filled with oils and attars of every type: jasmine, champaca, rose, kewda, three kinds of lotus, ginger lily, gardenia, frangipani, lavender, rosemary, wintergreen, geranium, and many more I've never heard of. The two bestsellers are *khus,* a cool, tranquil attar made from the roots of vetiver; and *hina-shamana,* a warm and woodsy compound said to be the favorite of Ghalib, a nineteenth-century Urdu poet of the Mughal era who was known as a ladies' man.

When we arrive, Mehrotra is busy with a young Muslim couple

---

* I recall "Ali Baba and the Forty Thieves" only vaguely. When I read the story later, I can see why it would be a point of pride; the leather oil vessels are crucial to the tale. In the story, the robber captain makes thirty-seven of his thieves hide in thirty-seven man-sized leather jars, cuts a slit in the top of each so they can breathe, and rubs the outside of the vessels with oil to authenticate the ruse. He disguises himself as an oil merchant and takes advantage of Ali Baba's kindness to spend the night at his house, planning to murder him in his sleep. The story's heroine slave figures it all out while her master Ali Baba sleeps, and gruesomely kills each of the thirty-seven bad guys in their hiding places by pouring boiling oil into the vessels.

who've traveled eighty miles to stock up on attars. Although his cus-
tomer base has contracted with the growth of mass-market perfumes,
many of his most faithful clients are Muslims, who use only traditional
attars for fragrance because Islam prohibits alcohol-based perfumes.

Shukla calls over a tea vendor selling *kulhad wali chai*. The man
juggles a tall stack of *kulhad* cups in his left hand and a steaming pot
of milky tea in his right, and still manages to pour with a flourishing
stream. I let my tea soak to capture the scent of the *kulhad*. I swirl it
around, then taste, smell, and drink it all at once.

By now, Mehrotra has fetched the *mitti attar*. It sits on the black
stone counter in an inch-tall glass bottle. I twist off the little gold cap,
close my eyes, and breathe in the scent of the Indian rain. It smells like
the earth. It smells like the parched clay doused with pond water in
the Siyarams' backyard. More than anything, it smells like the tea in
my little *kulhad*.

The aroma is entirely different from the memory of rain I carry
from my childhood and my part of the world—ozone-charged air, wet
moss, Thomas Wolfe's "clean but funky" scent of the South. But it
is entirely appealing: warm, organic, mineral-rich. It is the smell of
waiting, paid off: forty years or more for a sandalwood tree to grow its
fragrant heartwood; four months of hot, dust-blown summer in northern India before the monsoons arrive in July; a day for terra-cotta to
slow-fire in a kiln.

I ask Shukla, the supersmeller, to tell me what the scent brings
to his mind. "It is the smell of India," he says. "It reminds me of my
country."

I n my country in recent years, rain likewise has become a singular national reminiscence. But there is no earthiness about this American
aroma. No electrical zing of ozone. Nothing as dank as geosmin to
wrinkle the nose. With climate terrain as different as the Mojave Desert from the Louisiana swamps, it is not, as in India, patriotic nostalgia

for the smell of wet ground. It is, instead, a widespread recognition of the smell of wet laundry.

From cleaning supplies to beauty products, American retail shelves are awash in rain-themed products. In the household aisles: Refreshing Rain laundry detergent and dishwashing liquid, Renewing Rain fabric softener, Morning Rain carpet freshener. When it's time to scour the toilet, Clorox has you covered with a Rain Clean bowl scrubber.

As it freshens our clothes, counters, and commodes, we Americans also like rain-clean bodies. We shower with Pure Rain soap and Rainbath gel, soak in Midnight Rain bubble bath, and wash our hair with White Rain shampoo. Men can roll Granite Rain deodorant across their armpits while women rub the scent of Rain-Kissed Water Lily under theirs. The hyperhygienic woman can even freshen up with Tropical Rain–scented douche.

None of these scents existed in my 1970s childhood (though my favorite shampoo embodied the most blatant scent marketing ever: "Gee Your Hair Smells Terrific"). I tapped the International Fragrance Association of North America to help me figure out how rain ended up in so many U.S. product lines in the years since.

One of the top perfumers behind the craze turned up in Marietta, Georgia, just north of Atlanta. Heather Sims, a redhead with a refined southern accent, has a chemistry degree from the University of Georgia and the nose of a supersmeller. She's director of perfumery at Arylessence, a family-owned fragrance house that has become one of the largest U.S. developers of scents for beauty and personal-care products; laundry, cleaning, and home items like scented candles; prepared foods and teas; toothpastes, gum, lip gloss, and more. Arylessence's scientists churn up butter flavor for microwave popcorn, a Cajun kick for seafood boil, and subliminal air fresheners for hotels and retailers. They are the reason your Omni lobby smells of calming lemongrass, and the menswear retailer Tommy Bahama emanates a subtle piña colada—all in the name of marketing.

As a child growing up on a farm in rural Georgia, Sims often noticed scents in nature that other people did not—or detected them first, including a coming rainstorm. Her chemistry degree led her to Aryl-essence's fragrance lab right out of college. She arrived just as manu-facturers of cleaning supplies began to abandon the floral-dominated fragrances of the 1970s and '80s, responding to consumers who found them too reminiscent of Grandma's soap. At around the same time, most laundry detergents had become equally effective at conquer-ing dirt. Consumer choice was evolving to hinge on whether clothes seemed "fresh"—a notion that is wrapped nearly entirely in scent. Arylessence's marketing surveys show the smell of laundry detergent now drives nearly 80 percent of repeat buys.

Enter the rain. Despite geosmin and ozone, wet moss and mold, American consumers have what Sims describes as a fantasy of rain as fresh. This is not the case in damp regions such as Britain and Ireland, where products harkening clear skies and fresh air—"Breath of Fresh Air," "Blue Skies, "Sunshiny Days," "Sunshine Lemons"—far outnum-ber anything to do with water, be it rain, morning dew, or ocean spray. My Irish friend Susan Devane, who operates vacation cottages in the Wexford countryside and checked the store shelves there, could not resist adding her amusement: "It'd be a joke here. Honestly. People would just laugh: 'Ah Jaysus, ya wouldn't need to be addin' rain scent to the washin' today!'"

As an Irish consumer's dream of fresh is "Sunshiny Days," Sims says, an American's might begin with the childhood memory of a rain-storm and also trigger a fantasy of rain's bounty—rolling green hills, waterfalls, forests.

Chemists can't use gassy ozone to harken a coming storm. And they wouldn't dare add geosmin; no one wants the laundry to smell of fish. Instead, Sims and scent scientists conjuring rain might use anywhere between fifty and seventy-five different chemicals that call freshness to mind, starting with a wide array of compounds known as aldehydes. Aldehydes can be toxic, such as formaldehyde, and other-wise nasty, among the emissions from coal-fired power plants, forest fires, and diesel engines. But aldehydes with the longest carbon chains can produce strikingly fresh, aquatic scents. Many of these are behind the most famous perfumes in the world, including an unprecedented concentration in Chanel No. 5.

In fact, it was Gabrielle "Coco" Chanel, obsessed with cleanliness and highly sensitive to smells, who gave rise to the very notion behind the decline of India's attar industry. This was the spin that synthetic perfumes such as Chanel No. 5 are more refined than scents distilled from nature. Chanel was said to abhor simple flower fragrances she believed were used (unsuccessfully) to camouflage the odor of "unwashed" women.

"I want to give women an artificial perfume," she pronounced. "Yes, I really do mean artificial, like a dress, something that has been made. I don't want any rose or lily of the valley. I want a perfume that is a composition." Chanel hired Ernest Beaux, a celebrated French-Russian chemist and perfumer, to develop the formula that would become Chanel No. 5. Beaux said he chose the famous fragrance's set of aldehydes based on his scent memory of an expedition to the Arctic; he tried to re-create the redolence of Arctic lakes and rivers in the midnight sun.

Today, when Sims's clients seek to conjure rain, she also scans her aquatic memory and her periodic table. Following the aldehydes, she might reach for a synthetic muguet to re-create Coco Chanel's outcast lily of the valley. Next, she might add Calone 1951, a molecule developed by Pfizer in 1966 that imparts an exceptionally intense olfactory sea breeze. Water molecules alone have no smell, so lab-made rain scents, whether in detergent or perfume, often harken "blue notes"—sea and lake—and greens, fresh-cut grass, or clover.

"It is a fantasy," says Sims. "We are looking to create that fantasy of the cleanliness of rain." Ultimately, rain's essence wells up from your own fantasy, your childhood memory of rain, and where in the world you experience storms. It might be India's monsoons on parched earth, Helen Keller's hay meadow, or the warm whiff of steaming asphalt mixed with sweet frangipani—rain's balm in America's most tropical city.

———

ELEVEN

CITY RAINS

Miami is my favorite city of rain. If you've got to endure a deluge, it might as well be warm, fall amid fairy-tale banyan trees and Mediterranean architecture, and last less than an hour.

On the southernmost mainland, rain's applause rings loudest from the palm trees, fronds drumming with steady appreciation all through the wet season. Florida summer afternoons create just the atmospheric chaos a thunderstorm craves. As the sun warms the peninsula through the day, heat begins to rise off the land. Sea breezes swoop in from the Atlantic Ocean and the Gulf of Mexico. When these cool, damp drafts collide over the landmass, they push the warm air higher—sending huge, moisture-laden currents aloft. The blue-sky puff clouds of morning give way to silver Rubens by midday. Through the hot afternoon, the currents soar higher. Cerulean towers build and then vanish in a blue-black squall that dwarfs the beaches, marshes, and skylines of South Florida.

As the storm moves in, the sticky air turns cool. The thunder does not rumble in the distance so much as surprise with the closeness of its first sheet-metal clang. On the beaches, lifeguards blow their whistles and mothers wrestle wind-whipped blankets. In the Everglades, flocks of white ibis flash in the inky sky as they fly for cover in the hardwood hammocks. In Miami's Coconut Grove, wiry men selling mangoes on the highway medians make a break for the underpasses.

A few plump raindrops fall as five-second warnings before everything ordinary vanishes in an all-out tropical wash.

Rain gives Miami its substance as well as its tropical signature: South Florida's water supply depends entirely on rainfall to fill its aquifers because of what scientists call a hydrologic divide that severs the bottom of Florida from the bubbling springs and inflowing rivers that define the central and northern regions. Without rain, Miami would have no prehistoric-looking gumbo limbo trees, no hot pink bougainvillea pouring over walls and trellises, no bananas and plantains with their rain-slide leaves that ingeniously empty inward to growing trunks when the plants are small and outward to the root zone when they're mature.

Just as rain is the sustenance for tropical fruits, flowers, and trees, it is the balm for the oppressive summer heat of an overbuilt city. Miami's deluges cool off the asphalt streets and sun-beaten rooftops, so hot by afternoon the first drops sizzle to steam. They slow the breakneck traffic on Interstate 95. They sheet down the glass skyscrapers of Brickell Avenue to give workers in the financial district a taste of wild nature in their monolith offices. When the primordial storms move on, they leave Miami's pastel-colored buildings and frangipani trees glowing in a divine luminescence of sunlight streaming through dark clouds.

Ever the Janus, rain's intensity in South Florida can also leave burst sewage pipes that foul Biscayne Bay; streets turned to small rivers and parking lots to ponds; and soccer fields that resemble swamps. Like the rainmakers convinced they could blast storms from the sky or bring them on, the developers who dreamed up the cities of South Florida believed human ingenuity could keep capricious rain—too much or too little—in check. In fact, every inch of wetland or forest lost to every new inch of asphalt or concrete blocks rain's return to the aquifers or out to sea, making for an odd mix of human-made scarcity and flooding. Every razed mangrove makes the rising seas, storm surges, and torrential rains associated with climate change that much worse.

Miami is not unique in having developed in defiance of its hydrology. But unless the city can mend the mistake, it could be the first metropolis to succumb to it.

I n the summer of 1977, when I had just turned eleven, my mom and stepdad moved us away from the storms of Florida to our new home in Southern California. It happened that California was searing in the worst drought in its history; 1977 was the driest year ever recorded to that time, with less than half a normal year's rainfall. I don't remember those records, or whether I minded the dramatic change in my landscape, the glistening wet palm trees of South Florida replaced with skinny counterparts stretching from the boulevards of Los Angeles like tall straws in search of a drink. My most enduring memory is of the concrete channels, ubiquitous as the freeways, and how gargantuan they seemed for the tiny trickle of water they carried. Some of the kids in our neighborhood sneaked into the culverts to slide down the steep smooth walls on cardboard boxes, forbidden but as common as surreptitiously swimming in a Florida canal, just without the water.

Historians have variously described both California and Florida as lands of sunshine, states of dreams, and empires of water—the latter built by government engineers. They are also coastal testaments to rain's subjugation and ultimate sovereignty. It is rain's nature to erode mountains, carve canyons, swell rivers, and carry torrents of water, mud, and all else it catches to the sea. Cupping Los Angeles in every direction but the coast are mountain ranges including the mighty San Gabriels. From the Pacific Ocean to the tallest San Gabriel peak—nicknamed Mount Baldy—elevation soars from sea level to more than 10,000 feet. As atmospheric rivers roll off the Pacific in winter and meet the mountains, the clouds unload. In the eighteenth century, the native Tongva people tried to warn the Spanish not to build in the paths of the jumpy Santa Ana, San Gabriel, and Los Angeles rivers, which raced from the mountains to the valleys before erupting from underground passes into willow-lined channels and marshes of the coastal plain. In the nineteenth, flood-weary Mexicans tried to explain the same thing to Gold Rush–era Americans.

Every generation ignored the last. Crops, then rooftops, rose in the floodplains. L.A.'s first real estate boom rode in on the railways of the Southern Pacific in the 1880s; tracks were built alongside the wild rivers. A torrent in 1914, so fierce that alluvium from coastal farmland beached a steamer in Long Beach harbor, led to flood-control laws.

Emboldened by dams, developers built out the basin. The *Los Angeles Times* habitually scolded the rivers when bridges washed out. Until 1938, most Angelenos were still unaware of the soaking El Niño storms that cut atypically south to make landfall on the coast. In late February and early March of that year, five solid days of rain soused the already-saturated mountainsides. The dams were overwhelmed. Canyon washes burst their banks. Floodwaters rushed down creeks, washed out bridges, and surged into towns throughout Southern California. When the rains stopped, the mountains continued to disgorge. The rivers rose still higher. Nearly one hundred died across the region, including five members of a family in North Hollywood and three children from one Orange County family. In L.A. alone, more than 1,500 homes were lost and 3,700 residents sheltered by relief agencies. Southern Californians vowed to dry out and rebuild—and to do so right where they were, this time with a big assist of federal dollars. Until then, flood control had included capturing rainwater to supply local towns and farms. Now, imported water from the Sierra to the north and the Colorado River convinced the region to divert all its rain to sea and rely on someone else's. Over the next three decades, the U.S. Army Corps of Engineers built more than one hundred stadium-sized mountain debris basins, five mammoth valley flood-control dams, and 350 miles of concrete river channels. The Los Angeles River became a fifty-mile-long storm drain jutting ramrod across its old sinewy path. At the turn of the twentieth century, the river was still wild enough that grizzly bears lumbered down from the hills to feast on the steelhead. By the turn of the twenty-first, it was a massive concrete ditch, flowing most of the year with nothing but effluent from sewage-treatment plants upstream.

From California to Florida, the era of federal flood control—along with America's plumbing half the continent to pipe freshwater to arid cities—profoundly changed the human relationship with rain. Building against rain instead of with it had devastating consequences for the coasts. An estimated 85 percent of Los Angeles is urbanized, 65 percent of it paved over—sealed by impervious surfaces. Every subdivision and shopping mall, parking lot and pancake house prevents rain from soaking back into the ground. The rain that used to find its natural course to the aquifer or sea is now channeled, given a new name—"stormwater"—and poisoned as it rushes over dirty streets and down gutters.

In California and nationwide, this stormwater runoff has become the single largest source of pollution fouling beaches, major bays, and rivers. Like a cat that keeps dragging a dead mole back to the bedside, rain carries back to us the pollution and wastes we thought were out of sight. Running over all that asphalt and concrete, rain picks up toxic metals, oil and grease, pesticides and herbicides, feces, and every other impurity that can make its way to a gutter. Flood-control contrivances rush the stormwater off the streets and car parks, into the concrete channels, and back to the Pacific Ocean.

California health officials are challenged with how to warn non-English-speaking families to avoid eating any toxic fish that they catch. Elaborate charts warn that children should not eat sand bass, kingfish, barracuda, or croaker from the Palos Verdes Peninsula. After a rain, Southern California's surfers know to stay out of the water—no matter how good the waves may be. Those who defy the wisdom can end up with pinkeye, fever, or diarrhea. "The water will have this weird, funky smell to it," observes Sean Stanley, a twenty-six-year-old who has surfed L.A. County's beaches since childhood. "It's murky. You'll see soda cans and plastic bottles, oil from the cars. All the runoff from the city gets in there."

The story is familiar on the east and west sides of Florida's peninsular point. At the mouth of the St. Lucie River on the Atlantic side, and the Caloosahatchee at the Gulf, health officials routinely post advisories: "High bacteria levels. Avoid contact with the water. Increased risk of illness at this time." Before its farms and cities, southern Florida was dominated by the Everglades, a shallow mosaic of freshwater and sawgrass that flowed south 130 miles to the sea. The late writer and Everglades champion Marjory Stoneman Douglas often referred to the great marsh as South Florida's "rain machine." The Everglades absorbed the deluges, filled up the chain of lakes at the north end of the glades, topped off the groundwater in the aquifer, and sent a vast sheet flow southward, returning rain to the mangrove-lined estuaries of its birth.

To develop Miami and the dozen other major cities that hug the Atlantic Ocean in southeast Florida, boosters and government engineers drained, ditched, and diverted the great rain machine. They built legions of canals to irrigate farms in dry season and push the *devastating, ruining, havoc-wreaking* rains out to sea in wet times. Following the Mississippi River flood of 1927 and the 1928 hurricane that crashed Lake Okeechobee's earthen dike, drowning 2,500 farm workers south of the lake, the Army Corps began to build a massive levee around Lake O.'s south rim. After the 1947 wet season dumped 108 inches of rain in South Florida—more than double what falls in a normal year—Congress authorized the Central and Southern Florida project to harness every future drop and put an end to the cycle of too much or not enough. As with the Los Angeles River, the army engineers straightened the meandering, hundred-mile Kissimmee River, the Everglades headwater, into a fifty-six-mile "Dirty Ditch," as it became known for draining polluted runoff to Lake O. They built towering gates on the east and west sides of Lake O. to push water through canals to the Atlantic and Gulf when too much rain topped off the lake. The 1,000 miles of canals, 720 miles of levees, sixteen pump stations, two hundred gates, and other instruments of order were meant to calibrate the rain machine to human time and space. Instead, they sent it out of whack, turning the natural hydrological cycle into an artificial, hydro-*illogical* quagmire.

Draining and paving over South Florida brought a well-established record of devastation to water and wildlife. Now, despite multibillion-dollar restoration efforts, the engineers' order is turning against the human population, too.

Most of South Florida's 7.5 million residents live atop the broken rain machine; less than half the original Everglades remains. Without its sponge, the region is prone to severe flooding that can overflow canals and stall cars in swamped streets. But the most profound illogic is that South Florida regularly contends with drought emergencies even while surrounded by freshwater—trapped as it is in berms, canals, crop furrows, culverts, dams, dikes, ditches, pipes, plants, reservoirs, runnels, sluices, storm ponds, tanks, weirs, and wells. The fast-growing cities in the region overtapped the aquifer and now scramble to build costly drinking-water plants even as the Everglades plumbing flushes 1.7 billion gallons of rainwater to sea every single day.

On the flood side, when too much rain swells Lake Okeechobee, water managers must release billions more gallons to avoid a catastrophic breach of its dike. In the lake, the stormwater simmers into a toxic brew with agricultural and urban wastes including sewage, manure, and fertilizers. Dumped into the St. Lucie River on the east side and the Caloosahatchee on the west, the polluted water travels in black plumes down the rivers and out to estuaries at the Atlantic Ocean and Gulf of Mexico, where it can kill fish, spur toxic red tides, and shut down beaches.

In metropolitan Miami, the most immediate conundrum is how a finite and fragile tip of land can handle the sewage waste of 5 million people and nearly three times that many tourists. When I visited the Miami Dade Water and Sewer Department during rainy season, Virginia Walsh, a PhD hydrogeologist working on rainfall and climate models, explained how 300 million gallons of human waste flowing through the pipes on a normal day doubles on a day of hard rains, when stormwater overflows its drains and washes into the sewers. Miami is the world's metropolitan ground zero for sea-level rise—now contributing to flooding in Miami Beach even on sunny days. Walsh's computer models can tell city leaders where to shore up with bigger pipes, new pump stations, and gates that block backflow, all in the works. Scientists say Miami could face a rise of up to two feet by 2060—less than fifty years from now. The water may not wash over flood gates as we imagine, but seep up through the porous limestone and the storm drains, also elevating the consequences of Miami's exquisite rains.

For Walsh, sea-level rise is an engineering and urban-planning problem Miami will overcome, like the Netherlands and other low-lying urban areas of the world. What keeps her awake at night is not the rise that scientists are watching, but the potential for freakish rain events that they cannot predict. Throughout its history, Miami has seen years of extraordinary rain and years of drought; climate scientists say those extremes will worsen as the planet warms. "Sea-level rise we can deal with—until or unless we get to a point when we won't be living here anyway," Walsh told me. "What is going to have more immediate impact on people's daily lives is the rain."

If one place in America has lived through the worst consequences of its urban rain mistakes and begun in earnest to undo them, it is the erstwhile misnamed City of Rain, Seattle. A jolting human tragedy, along with the near-ruin of Seattle's flashing silver amulet—the salmon that swim through culture as much as nature in the Pacific Northwest—helped the larger community embrace what began as a grassroots effort to mend the city's relationship with its iconic rain.

When I visited in winter, a leaden monotony hung over the central

business district, bringing dispirit to downtown. Contrary to reputation, Seattle's urban pallor is not born of rain, which falls almost im-
perceptibly from silvery clouds that match the nearby waters of Puget
Sound. More than cinereal landscape, the gloom rises from the cement
hardscape. The busy streets are paved dark gray, the wide sidewalks beside them light gray. The skyscrapers rise in shades of gray. The hulking
freeways, ramps, and overpasses: gray. The monorail track and its ele-
phantine pillars: gray.

Trudge the ashen sidewalks northwest to Seattle's Belltown neighborhood, hang a left on Vine Street toward the sound, and a ten-foot-
tall, bright blue rain tank pops out in the dullness, tipped in whimsy
toward a red brick building. Atop the tank, green pipes in the shape
of fingers and a thumb reach out, the stretched index finger connected
to a downspout from the building's rooftop. Rainwater flows from roof
to finger to tank. From the thumb, the rain pours into a series of descending basins built between the sidewalk and the street. The basins,
in turn, cascade to landscaped wedges growing thick with woodland
plants. For two blocks as Vine Street slopes to the sound, the rainwater trickles down a runnel and through street-side planters, shining
stones, and stepped terraces, enlivening the roadway with greenery,
public sculpture, and the sounds of falling water.

The project, called Growing Vine Street, began as a small effort
among residents and property owners to turn their stretch of a former industrial neighborhood into an urban watershed. Twenty years
later, such street-greening projects have become a key part of the city's
strategy for managing stormwater, the major pollutant fouling Puget
Sound. This rain runoff impairs virtually every urban creek, stream,
and river in Washington; makes Pacific killer whales some of the most
PCB-contaminated mammals on the planet; is driving two species of
salmon extinct; and kills a high percentage of healthy coho salmon
within hours of their swimming into Seattle's creeks, before they've
had a chance to spawn.

Pollution is only one side of stormwater's twin troubles. Flooding is
the other. In December 2006, extraordinarily hard rains fell on parts

of Seattle in what became known as the Hanukkah Eve windstorm. Flooding and mudslides closed major roads. Cars floated on Mercer Street under the Aurora Avenue overpass. Thousands of people driving to a Seahawks game were stranded in their cars for hours. Sewage surged up and out of toilets and drains. But the greatest tragedy occurred in a windowless basement in the Madison Valley neighborhood. A well-known voice actress named Kate Fleming ran her audio-production company from her home at a Mercer Street crossroads where the sewage pipes were not large enough to handle the runoff from this big a storm. As the water rose in Fleming's neighborhood, then began to fill her basement, she ran down to try to save her recording equipment. A surge of floodwater followed, and trapped her inside. By the time firefighters cut a hole through the floor above her and pulled her out, it was too late. Fleming died at a nearby hospital, at age forty-one.

In Seattle and all the cement-suffocated cities of the world, restoring hydrology can reduce harm. Seattle is at the forefront of a rain revolution that gives floodwaters more natural places to drain, replaces impervious surfaces with porous ones, and clears rain's pathways to aquifer and sea. Like many revolutions, this one began in the streets—actually along the edges, as engineers and landscape architects replaced concrete curbs and drains with grassy swales, and planted hundreds of trees and shrubs to help filter and slow the flow of stormwater. The first green street they finished, in 2001, eliminated nearly all runoff.

Rain gardens are another of the strategies that prove as effective as they are beautiful. Washington State University scientists found that street-side gardens clean up 90 percent or more of the pollutants flowing through on their way to Puget Sound. Green roofs likewise absorb and clean up rain. The thousands of square miles of asphalt, black tar, and gravel that cook on urban rooftops aggravate flooding and stormwater pollution as they sheet dirty rain onto cities. Green roofs—like the inspired native prairie grasses and wildflowers that bloom yellow, white, and purple atop Chicago City Hall—can cut runoff by more than half. Cisterns like the blue one on Seattle's Vine Street capture the rain and store it for irrigation. Made by the public artist Buster Simpson, *Beckoning Cistern* and its reaching hand suggest Michelangelo's *Creation of Adam*. The idea is a new ethic for the way we live with water and rain. All the attention has helped people and businesses use less as they pollute less; total water use has plummeted since the mid-1970s even while population has soared.

Bigger picture, Seattle has spent tens of millions of dollars acquiring flood-prone blocks and transforming them into lush green spaces that hold stormwater during hard rains. Along a path at the new Madison Valley Stormwater Project, an eight-foot-tall stone sculpture is dedicated to Kate Fleming, bearing the words she always recited to herself before appearing on stage: "Be a light. Be a flame. Be a beacon."

Visiting Seattle's rain gardens, green streets, green roofs, stormwater parks, funky cisterns, and rain art installations, including a mesmerizing display of public rain drums that thrum to the beat of the skies, I could see it deserved the Rain City nickname even if it's not the rainiest place in the nation. Seattle is wet proof that we humans can come to live in harmony with the rain.

In the same San Gabriel foothills outpost where Charles Hatfield built his rain derrick and promised to quench Los Angeles in the spring of 1905, and only four miles from Frank Lloyd Wright's La Miniatura of 1923, a more modern rain lover set out to see if one soul among L.A.'s ten million, living in one single-family home among two million, could restore the natural path of rain on her small part of the terra firma.

Emily Green is a native Californian whose mother's family grew citrus in the San Gabriel Valley. In the great dam-building era of the 1920s, her grandparents attended openings of local dams and reservoirs to celebrate the water for their groves. When Green was young, her family built a vacation house in the British Virgin Islands whose roof fed an enormous basement cistern. "When it ran out of water, we ran out of water." Green became an environmental journalist (chanceofrain.com) and impassioned gardener—two pursuits that have her constantly scrutinizing the skies. Her pluvial ancestry seems part rain goddess, part Daniel Defoe fact finder: Green can describe the tousled romance of native grasses, or spend fifteen months investigating the L.A. Flood Control System to expose engineers up a creek—having run out of space for the mountain debris choking their dams, yet having left no rivers intact for sediment to flow down.

Green's modernist house in Altadena, just a few blocks from Angeles National Forest, sits toward the front of a 14,000-square-foot lot. In the back, she tends a large native garden and a small orchard of venerable fruit trees older than the house—oranges, lemons, tangerines, and an enormous avocado. Having already torn out 10,000 square feet of lawn, and the "fuck-you hedge" that lined the front yard like a castle wall when she moved in, Green fretted about using L.A.'s imported water for the citrus—as the rain that fell upon her house and landscape washed to the storm drains. Like a miniature of the city itself, her home was designed so that any rain hitting the roof or lot would flow to graded pavement encircling the foundation, then drain *away* from her garden to paved paths and an asphalt driveway, then pour into the street gutters, which in turn run to the great concrete culverts.

Green wanted to break the rain regime. In 2010, she hired guys with mallets to come bust out a concrete patio out back and the asphalt driveway out front. In 2011, she began her search for a rain-gutter installer with a sense of style, not an easy task. She found hers in a sheet-metal artist named Ruben Ruiz.

I have Green's hand sketch of her roof, with its calculations of how much rain she could collect from each section and her dream of metal flower sculptures that would chime visual and aural melodies with every shower. Depending on the grade of metal, rain tapping a tin roof or running in a gutter can create in children who grow up with it a calm contentment that stays with them all their lives—or a racket to send occupants to insomnia. Green chose a decent grade of galvanized steel. It calmed.

When I visited Green in Altadena, Ruiz had just finished installing the clean-lined steel gutters. They fit the gray-tiled roof as handsomely as a well-tailored jacket—none of the ugly downspouts, awkward angles, or corrugated aluminum pieces that often make rain gutters an architectural crime. In front of the house, lengths of chain lead from small drains to the flower rain-catchers, where a metal butterfly flits at the tip.

It was all so idyllic it was hard to remember we were standing five miles from the freeway and fifteen from downtown Los Angeles. But soon enough, the vulnerability of Green's sustainable rain plan would be revealed. She hit a stumbling block bigger than L.A.'s entire Flood Control System. It stopped raining.

In spring 2014, California's reservoirs dropped to less than half their capacity, revealing vast deserts of cracked mud and in one case the ruins of an entire Gold Rush mining town. The California State Water Project—the labyrinthine conveyances that move billions of gallons to millions of residents and farm acres—announced for the first time in its history that it would hold back water to cities and farms. Seventeen communities faced running out of drinking water. Farmers had to fallow a half million acres of land. California's remaining salmon rivers had so little flow that wildlife officials were figuring out how to truck young fish from hatcheries to sea. And snowpack in the Sierra Nevada that would normally rehydrate all of it was just 15 percent of the average. Without some rain, the dream state was headed for a nightmare.

After consecutive years of below-normal rainfall beginning in 2011, California's drought became the worst since 1977—and then the worst in history. The parallels to the drought of my childhood were many. Governor Jerry Brown, the nation's youngest governor then, was back in Sacramento, now the nation's oldest governor. He was trying to push through some of the same solutions he had nearly forty years before, such as a canal to capture freshwater from a tributary of the fragile Sacramento–San Joaquin Delta in the north and move it south.

And many Californians, blasé about the emergency, continued to water their turf grass. During a rare deluge in February, the Los Angeles Department of Water and Power had to send out a press release urging customers to shut off their outdoor sprinklers so they wouldn't waste water on already soaked lawns and gardens.

Still, contrary to the Los Angeles clichés of concrete culture, plastic people, and dystopian future, the most urbanized arid city in America is full of revolutionaries like Green who have worked for decades to bring the city in harmony with water and its rainfall. Angelenos use far less water (about 152 gallons a day per person) than their counterparts in the state capital of Sacramento in the north or Palm Springs to the south, where residents average 736 gallons a day to fill pools and irrigate mega-lawns in the desert. The city's lawns are slowly transitioning to native plants and food crops. Hundreds of acres of community gardens have spread across Los Angeles, which also has more farmers' markets than any other American city.

At Woodbury University in Burbank, the architect spouses Hadley and Peter Arnold run the Arid Lands Institute, devoted to reversing the twentieth-century course of channelizing rivers and importing water long distances as local rain flushes to sea. (The couple met as graduate students at the Southern California Institute of Architecture in Los Angeles, when he took to feeding scraps to her Welsh corgi, auspiciously named Splash.) The Arnolds see L.A. as the great western test for what they call drylands design. The idea is to make rain a centerpiece of architecture, building codes, and zoning laws rather than a scourge. They are looking to the successful storage and land-use strategies of the past, from the sharing model of nineteenth-century Mormon irrigation districts in Utah to the extensive cistern works of the Roman Empire and North Africa.* "Cities got water right when they had to think like a drop of rain on the path of gravity," Hadley

---

* The cisterns of ancient Carthage in Tunisia are celebrated as Roman, but the Punic people built cisterns throughout their lands before the Romans ever arrived.

Arnold told me. The water and power-grid feats that allowed western water to be piped long distance created not only the obvious costs to rivers like the Colorado but invisible ones—not least of which the carbon emissions now altering the climate.

Tucson, Arizona, is the quintessential arid city changing to embrace its manna from the desert sky. Rain gutters gleam on many homes and businesses; copper downspouts are a status symbol. Rain tanks are becoming as ubiquitous as in the Caribbean, where rain is still the sole source of water on some islands. Bolstered by a cadre of young conservation-design gurus now training others around the world, Tucson is tearing up asphalt spillways and non-native landscaping so that floodwaters from the North American monsoons flow to natural basins, berms, and desert gardens. While most cities deplete their groundwater, Tucson is returning more to the aquifer than it draws. Still, it is hard to see how the oasis now bursting with a million residents will thrive in a rain-limited future. Tucson's population is expected to double by 2050. If we do not lower the carbon emissions that are warming the Earth, scientists predict the arid Southwest between the Texas Panhandle and Southern California could lose up to half its rainfall by century's end.

By that time, Los Angeles hopes to have reversed the flush of rain to the Pacific Ocean. The city has launched an ambitious retrofit of the entire stormwater system, including massive spreading grounds on those publicly owned lands that absorb the most rain, and incentives for infiltration on private property, distributed like the goal of rooftop solar. "When you think of the scale, we can do 1,000 neighborhood infiltration basins, we can do 100,000 rain gardens," says Mark Hanna, the L.A. water engineer who is overseeing the plan. "It's an enormous amount of water saved."

I asked Hanna the obvious: "What if it doesn't rain?" His answer was optimistic. I could not help but think of the great rainmaker Charles Hatfield, and his prescient understanding of how California drought is often followed by deluge. "It does rain and it will rain,"

Hanna told me. "And when it does, unless things go really strange on us, the rain will tend to fall in the mountains and gather in the canyons and accumulate in the low spots. As long as gravity doesn't go anywhere and rain remains wet, it will go to these places. We need to capture every drop that falls."

In the last piece of L.A.'s water puzzle, a tireless grassroots effort to restore the L.A. River finally won the support of mayors, and then the Army Corps that turned it into a culvert in the first place. In 2014, the Corps approved a billion-dollar plan that will tear out miles of concrete, widen the river, and restore hundreds of acres of wetlands and native willows in the middle of the city.

Amid the thrill of a real river coursing through L.A., Hadley Arnold hopes Angelenos will push restoration to a new level—to encompass all the water in the hydrological cycle including the atmosphere. "The work on the river is incomplete without also working on the water that falls in the foothills, in the low-density residential fringes, in the high-density core, in the commercial and industrial areas, in the airport," she says. "The idea is to see water not in a 54-mile line, but in a field."

As South Florida learned, river restoration alone won't be enough without a widespread shift in how Angelenos treat their water and their rain. In the Everglades, federal and state engineers have spent fifteen years to restore the Kissimmee River and its floodplain, dynamiting water-control structures, remaking meanders and oxbows, and reestablishing 12,000 acres of wetlands. Native aquatic plants are thriving. Largemouth bass and sunfish returned. Snowy egrets, blue herons, and other wading birds that had vanished flocked back in astounding numbers, in some cases more than double what scientists expected. But the victory is hollow for the Floridians who've watched seagrass, fish, dolphins, and manatees die along their coasts. Lake Okeechobee's floodwaters still flow in eutrophic plumes to the estuaries. The farms and the urbanites of South Florida still send off a steady stream of pesticides, fertilizers, manure, and sewage, all brought back 'round by the dutiful rain.

In L.A., Emily Green was similarly discouraged. After she'd run through her savings to build the metal gutters and flowers that never clinked with rain, her city water use—and her water bill—more than doubled as she irrigated to keep the old citrus grove alive through the drought. She was standing in line at Home Depot when it sank in that home owners with rain gardens would not be enough to lead the arid metropolis to a rain revolution. In the middle of the drought, Angelenos stood with shopping carts full of weed killers, grass killers, insect killers, weed preventers, lawn fertilizers, and lawn sprayers, dusters, and wands—hundreds of products devoted to keeping the lawn crowned king of L.A. Without rain's dilution, the chemicals would concentrate in gardens, street gutters, and storm drains, becoming more toxic before making their way to the sea.

The last time I talked to Green, she was agitating about all the public green lawns of L.A., including the one that fronts City Hall. She was also tackling a new challenge, this time to see how one Angeleno could help change car culture. She sold her Prius and bought a bright orange bicycle at Steve's Bike Shop in Altadena, with an electric assist to cope with the steep San Gabriel foothills.

With her new mode of transportation, I figured the skies over Los Angeles would finally begin to pour.

V

MERCURIAL

RAIN

TWELVE

STRANGE RAIN

On a June day in 1954, Sylvia Mowday had taken her children to a park in Sutton Coldfield, just north of Birmingham, England, when the overcast skies darkened and a rainstorm caught them by surprise. Mother, son, and daughter began to run for shelter, but gentle thuds against their umbrellas froze them like statues. Something was falling with the rain. It was too soft for hail, and seemed to be alive. The family soon realized they were in a tempest of tiny frogs. Looking to the sky, they watched wee frog bodies fall with the individual symmetry of snowflakes. Mrs. Mowday estimated that thousands of frogs fell over several minutes. Afterward, they "were afraid to move in case we trod on them."

Throughout history, bewildered observers have sworn to similar episodes of frogs falling with rain. They show up in Greek literature, in the work of chroniclers from the Middle Ages, in accounts from French soldiers fighting the Austrians at Lalain in 1794: A hot afternoon was broken by such heavy showers that 150 soldiers had to abandon their trench as it filled with rainwater. In the middle of the storm, tiny toads began to pelt down and jump about in all directions. When the rain let up, the solders discovered more toads in the folds of their three-cornered hats.

Frog and toad rains, fish rains, and colored rains—most often, red, yellow, or black—are among the most common accounts of strange rain, reported since ancient times. "It has very often rained fishes," wrote the Greek historian Athenaeus in A.D. 200, in his *Deipnosophistae,* or the Banquet of the Sophists. He went on to recount tales of fish rain—including one scaly downpour that went on for three days—and falling frogs. He recites the oldest account of frog rain from a book of history (now lost) written in the second century B.C. by the Greek philosopher Heraclides Lembus. So great was the frog-fall that it poisoned the wells, forcing people to abandon their homes:

> *In Paeonia and Dardania, it has, they say, before now rained frogs; and so great has been the number of these frogs that the houses and the roads have been full of them; and at first, for some days, the inhabitants, endeavouring to kill them, and shutting up their houses, endured the pest; but when they did no good, but found that all their vessels were filled with them, and the frogs were found to be boiled up and roasted with everything they ate, and when besides all this, they could not make use of any water, nor put their feet on the ground for the heaps of frogs that were everywhere, and were annoyed also by the smell of those that died, they fled the country.*

The account calls Exodus to mind; frogs are one of the ten plagues God sends down to Egypt, where they swarm the houses, bedchambers, beds, people, ovens, and kneading-troughs. In 1946, the professional skeptic Bergen Evans—later arbiter for the television series *The $64,000 Question*—proposed that stories of falling frogs and fish "are a sort of detritus of the old belief in spontaneous generation," with roots in ancient myths and biblical references to aerial waters "above the firmament." But frog and fish rains show up far too often to be rejected as what Evans called meteorological myth.

In 1873, *Scientific American* ran eyewitness accounts of a storm that rained frogs on Kansas City, Missouri. In 1901, witnesses in Minneapolis swore to a similar slimy deluge. Reports of frog rain continue in modern times, although less frequently than in the past. Residents of

Naphlion in southern Greece were surprised by a rain of small green frogs in May 1981. A Belgrade newspaper reported a thick rain of frogs in the Serbian village of Odzaci in 2005. A villager, Caja Jovanovic, said he was watching a strange-shaped cloud when "frogs started to fall. I thought that a plane carrying a cargo of frogs had exploded." In 2010, hapless frogs fell upon shoppers in Rákóczifalva, Hungary, during a thunderstorm. The same year, frogs and fish rained from a cloud during a downpour in Nakuru, in Kenya's Rift Valley.

In the early twentieth century, Eugene Willis Gudger, an ichthyologist with the American Museum of Natural History and editor of the museum's *Bibliography of Fishes,* reported that he had authenticated seventy-one accounts of fish rain, spanning A.D. 300 through the 1920s. "I have personally never been so fortunate as to experience or even witness such a rain," he wrote, "but I cannot disregard the evidence recorded by scientific men." In one of his eyewitness accounts, from May 1900, family members in Providence, Rhode Island, were pelted with "squirming perch and bull-pouts, from two to four and a half inches long, which fell on yards and streets—covering about a quarter of an acre." A reporter from the *Providence Journal* gathered a bucketful.

Gudger believed there could be but one explanation: "High winds, particularly whirlwinds, pick up water, fishes and all, and carry them inland where, when the velocity of air and clouds becomes relatively lowered, the fishes fall to earth." He wrote that no one who had "experienced or even seen the prodigious effects and carrying power of a land tornado can have any doubt of the ability of a waterspout, a water tornado, to bring about a 'Rain of Fishes.'"

Still more peculiar rains reported over history have included hay, snakes, maggots, seeds, nuts, stones, and shredded meat (that last one is suspected to have dropped from a boisterous flock of feeding vultures). Dirt-toting rains are more common. In 1902, a massive dust storm kicked up in Illinois blew to the Eastern Seaboard, where it met up with thick rain clouds over New York, New Jersey, Connecticut, and Pennsylvania. The resulting strange rain was short but shocking for those caught in it: The sky rained mud. "People who were on the street were covered with mud spots," wrote a *New York Times* correspondent from the Finger Lakes region of New York. "Clothes hanging on the line were smeared." In Aurora, New York, the Reverend George P. Sewell wrote that the storm front was forty miles across and "discolored or soiled everything exposed to it."

In recent years, in certain parts of the world including Australia's Lajamanu in the Northern Territory and the village of Yoro in Honduras, people who remember fish rains from childhood have experienced them again as adults, grabbing buckets to collect dinner from heaven. Yoro has begun an annual carnival, Festival de la Lluvia de Peces, to celebrate the phenomenon. Australian scientists have some of the best long-term data on fish falls, beginning in the 1920s, but not enough to explain them definitively. Modern meteorologists agree with Dr. Gudger's theory that tornadoes and waterspouts are the most likely culprits. In one of the few reports of multiple species falling from the sky, in June 1957, thousands of small fish, frogs, and crawfish rained down on an Alabama town called Magnolia Terminal. A tornado reported around fifteen miles to the south was likely responsible, speculates the severe-weather guru Dr. Greg Forbes. But he and other scientists acknowledge it doesn't explain why an air current would pick up only small frogs or fish and not every other algae and creature from the same pond.

The strange rain of Labor Day 1969 in Punta Gorda, Florida, especially begs the question. During an otherwise normal rainstorm, golf balls seemed to hail onto the rooftops and roads; at least that's what local police lieutenant Clarence Walter claimed to the *St. Petersburg Times*, which ran the story under the header "Streets, Gutters a Duffer's Dream." The newspaper reported that "dozens and dozens and dozens" of golf balls lined the sidewalks, streets, and gutters following the rain, but gave no theories why. *Popular Mechanics* magazine speculated that a waterspout could have sipped up a ball-filled pond, then dropped its catch on the town. It seemed like an appropriate strange rain for golf-enthused southwest Florida. Often, the rain that falls upon us is simply returning what we've released to the earth.

J udging by sheer number of words published on the subject, no one person has given more thought to strange rain than Charles Hoy Fort. He was born in Albany, New York, in 1874, the oldest son in a wealthy family of grocers. As a boy, when Fort expressed far more interest in natural history than the family business, his father tried to beat shopkeeping into him. Instead, Fort became ever more rebellious and questioning, and eventually left home to make his way as a writer.

In the early 1900s he tried science fiction, including a novel about beings on Mars controlling life on Earth. When his fiction did not sell, Fort began work on what his biographer Jim Steinmeyer calls "the first book of oddities." Fort's 1919 *The Book of the Damned* was forerunner to popular collections such as *The Guinness Book of World Records* (first published in 1955), Ripley's, and National Geographic's *Weird but True* series. Those quirky lists are irresistible to kids. But Fort's work was not kids' stuff. He was out to needle the establishment—not only religion and philosophy but also science, which he claimed was a sham for excluding bizarre facts too messy to fit its theories. Those included frog and fish falls, mysterious lights or airships reported in the sky before airships were invented, and perhaps his greatest obsession, colored rains—red rains, yellow rains, "rain so black as to be described as a 'shower of ink.'"

Fort spent years digging up reports of strange rain, combing through archives in the New York Public Library and the British Museum. He ultimately collected some 60,000 newspaper clippings about these and other unusual occurrences, describing them as "damned" because, without a logical explanation, scientists tended to dismiss them. He would meticulously source his weird findings like a good researcher, carefully citing dates and publications that corroborated rains of toads,

frogs, snakes, eels, spiders, stones, pebbles, salt, cinders, coal, and ge-
latinous goo. But then he would ridicule the official theory and toss in
some wild speculation from the sci-fi-writer side of his brain: Perhaps
there was an invisible "Super Sargasso Sea" overhead, he would declare, a dimensional crossroads where things suddenly materialize or
disappear: "derelicts, rubbish, old cargoes from inter-planetary wrecks;
things cast out into what is called space by convulsions of other planets."

Many of Fort's pet phenomena, including frog and fish falls and
colored rains, have since been accepted by science—if not entirely explained. In 1981, U.S. Secretary of State Alexander Haig cited yellow rain as evidence when he accused the Soviet Union of supplying
chemical weapons to Communist Vietnam and Laos to use against
the Hmong people, in violation of the Geneva Protocol and 1972 Biological Weapons Convention. The biologist and bee expert Thomas
D. Seeley, now at Cornell, thought the description sounded like the
massive "defecation flights" of honeybees known to send bee poop and
pollen in rainlike showers of yellow. He later worked with the Harvard
chemical-weapons expert Matthew Meselson to denounce the government's claim and show "physical and biological evidence that yellow
rain is the feces of Southeast Asian honeybees." Refugee workers and
Hmong people who survived horrific experiences regardless of whether
chemical attacks occurred believe they did occur. Other scientists and
former CIA agents remain divided or uncertain. Fort could have told
them he knew of reports of yellow rain in history stretching back to
1695 in Ireland.

Fort likewise collected many reports of red rain. He was incensed
by the scientific explanations that they must be associated with Saha-
ran sandstorms. "My own impositivist acceptances are: That some red
rains are colored by sands from the Sahara desert; Some by sands from
other terrestrial sources; Some by sands from other worlds, or from
their deserts—also from aerial regions too indefinite or amorphous to
be thought of as 'worlds' or planets."

Falls of red rain and red dust are well documented today. Meteorologists do link most of them to the great Sahara. Weather satellites show the dust sweeping for thousands of miles into the Atlantic Ocean, gusting north to call children to write "Wash Me" on the cars of Wales, or blowing south, where it can weaken formation of hurricanes that otherwise might wallop the eastern United States.

In at least one intriguing case, though, blood-red rains have been colored by something more mysterious. For more than a century along the southwest coast of India in the state of Kerala, people have observed red rains so rich they can stain white clothes pink. Conventional wisdom had it that Kerala's red rains were caused by dust from a distant desert. But after a red rain in the summer of 2001, when researchers analyzed scarlet rainwater collected around Kerala, they found it contained no dust. It was full of microscopic red particles that looked like biological cells. Other scientists suspected they were spores from abundant algae that cover trees in the region. But physicists Godfrey Louis and Santhosh Kumar found that the pigmented particles (stored in laboratories, they maintain their deep red color these years later) did not contain the flagellae usually found in algae cells; other scientists have also disproved the algal-bloom theory.

Louis and Kumar hypothesized that the 2001 rain was linked to a meteor airburst that occurred over Kottoyam the day the red drops began to fall. The scientists caused a stir when they wondered if the red particles could be extraterrestrial in origin. Their latest study has been published with a team of astrobiologists and molecular biologists in the U.K. who were initially skeptical of the claims. The team found that the red rain cells survive and grow after incubating for two hours at 121 degrees Celsius (250 degrees Fahrenheit); most forms of life on Earth grow in a temperature range between 10 and 45 degrees Celsius. When exposed to the extreme heat, the red cells began to produce daughter cells. Additionally, the team reported that the fluorescent luminosity of the red cells bears "remarkable correspondence with the extended red emission observed in the Red Rectangle planetary nebula and other galactic and extragalactic dust clouds."

The scientists suggested an extraterrestrial origin. Others remain skeptical, but this much is certain: Charles Fort at last would have had a scientific theory he could buy into.

At the University of Oklahoma in Norman, environmental engineering professor David Sabatini always starts off his "Introduction to Water" class by asking students: "What's the purest water on Earth?" Rain is usually the most popular response, followed by mountain streams and snow. The answer is distilled water; made for microchips and not to drink. (The human body needs all the ions that water picks up from the earth, and our taste buds like them, too.) Like Isaac Newton unweaving the rainbow, Sabatini methodically debunks the students' responses, beginning with the rain.

Just as rain can reflect the joy or melancholy of the person caught in the downpour, rain's quality or cleanliness reflects the air and oceans it travels through. While many of us revel in the scent of rain from childhood, rain can also take on the sickening tastes and odors of pollution—or death. Survivors of the atomic attacks on Hiroshima and Nagasaki in 1945 reported black rains that poured for up to seven hours after the bombings. Some children wandering in the hellish aftermath were so thirsty they collected the inky rain and drank it, succumbing quickly to death.

Surviving prisoners of the Auschwitz death camp in Poland describe rains of unimaginable misery, stinking of diarrhea. It would take survivors years to remember rain's pleasantness, in the same way they had to relearn to use toothbrushes, eating utensils, and toilet paper—in the same way they had to relearn to smile.

Charles Hoy Fort was especially interested in black rains, which frequently fell on the British Isles in the nineteenth century. Shepherds on the upland moors gave the name "moorgrime" to the inky rains and soot that accumulated on the fleeces of their sheep. It was some of the first proof that industrial emissions could be carried for long distances.

It was also a reminder that what goes up must come down; sometimes, with the air-scrubbing rain.

Some of the tales of black rains of ash and pulverized pumice that Fort collected from Europe could be attributed to Mount Vesuvius in Italy. The volcano erupted five times in the second half of the nineteenth century. But chemists linked the greasy rains that could turn a white sheep black to the soot emanating from the manufacturing cities of northern England and central Scotland—then cranking out Mr. Macintosh's double textures and other textiles, along with a choking grime that blackened the air in London, in Manchester, and beyond. Simply put: The more pollution we pump into the atmosphere, the dirtier the rain.

In 1853, Charles Dickens opened his novel *Bleak House* with a soft black drizzle, "flakes of soot in it as big as full-grown snow-flakes," and "fog everywhere." Fog is essentially an eye-level cloud. If cooling air becomes saturated with water vapor, some of the vapor condenses around microscopic bits, forming droplets, which turn to fog. Given the exact same amount of water vapor, fog over a dirty city will be much thicker than fog over the sea because it gloms on to all those particles of smoke.

By Dickens's time, it was obvious that the atmosphere over industrial cities was no longer entirely natural. In the countryside, he wrote, "fog was grey, whereas in London it was, at about the boundary line, dark yellow, and a little within it brown, and then browner, and then browner, until at the heart of the City . . . it was rusty-black."

Londoners called the thick brews of smoke and fog that could alight on the city "pea-soupers" or "London Particulars." When a pea-souper claimed 1,150 lives at the turn of the twentieth century, a physician, Harold Des Voeux of the Coal Smoke Abatement Society, dubbed the choking air "smog." The region's coal-fired factory kilns and boilers, along with steam locomotives and ships, fumes from lorries and buses, and the soft coal that burned in the hearth of every home, would culminate famously in London's Great Fog of 1952. The worst pea-souper in London's history, it was bleaker than anything Dickens could have dreamed up.

As the calendar turned to December, a cold, damp fog settled on the city, absorbing all the smoke and soot in the air. Those conditions were not unusual. But on December 5, the winds stopped blowing. A band of warm, high pressure moved in, which trapped the colder air below and held the toxic smog in place, suffocating the city.

The smog thickened and darkened until visibility was reduced to near zero. Streetlights burned all day. People made face masks for themselves and their children before venturing outside. Once out, greasy soot rained upon them, and many could not find their way home. They inched along sidewalks by feeling the walls of buildings. They abandoned their cars. Bus and trolley services were halted, Heathrow Airport and the Port of London shuttered. People, press, and politicians were so focused on the fog's unusual consequences for daily life— criminals were exploiting it to conceal a rash of "burglaries, attacks, and robberies" and all football games were canceled, along with *La Traviata* when the murk filled Sadler's Wells Theatre—they were slow to recognize the unfolding human health disaster.

Ultimately, the Great Fog killed 12,000 people—4,000 over the five days it lingered, and an estimated 8,000 more in the months to follow—making it the worst peacetime catastrophe in British history. For many months afterward, the government tried to minimize the significance and magnitude of the deaths, and to characterize the killer fog as a natural disaster. Facing intense public pressure for action, the minister of health complained, "Anyone would think fog had only started in London since I became minister." It would take four years, but Parliament finally passed a Clean Air Act in 1956 that created zones where only smokeless fuels could be burned, and required relocation of power stations away from cities. The United States passed its Air Pollution Control Act in 1955, seven years after the same kind of atmospheric inversion killed twenty and sickened thousands in the mill town of Donora, Pennsylvania. But the U.S.

legislation and 1963 update called for nothing but research. Not until the Clean Air Act of 1970 did the federal government step in to regulate air pollution.

Great Britain has never again seen the black rains of its dark industrial history, although in the United States, a memorable black rain fell in South Boston in 1960. Engineers at the Metropolitan Transit Authority power plant combined pulverized coal and oil, creating a disastrous mess of emissions that mixed with the falling rain to produce a "black ink that would not wash off surfaces and foamed when it hit the street." Black rains and black snow also fell in the northern mountain ranges of India's Jammu and Kashmir in the winter of 1991; with no heavy industry in the region, scientists linked them to the burning of oil wells in Kuwait during the Gulf War. In the twenty-first century, bloggers in China have reported black rains associated with plants burning coal and poor-quality heavy oils on several occasions in Shenzhen, just north of Hong Kong. Residents reported that the rains can carry a pungent odor, cause a burning sensation to skin, corrode car paint, and leave raindrop-sized holes in flower petals.

Surely no strange rain could be as insidious as toxic black streaks falling from polluted skies. But there did turn out to be one: fresh, clean-looking rain that would also prove poisonous.

I t comes as no surprise that acid rain was first discovered in the gritty skies over Manchester. Robert Angus Smith, an English chemist, was Britain's first alkali inspector (aka air-pollution watchdog). In 1852, he discovered a link between Manchester's soot pollution and high levels of acidity he found in rainfall. Twenty years later, Smith described the problem of "acid rain" in a six-hundred-page book on the subject. For nearly a century, no one paid a bit of attention. Acid rain has no discernible taste or smell, no dramatic calling card of any kind. Its harm to ecosystems, along with the statues and buildings of antiquity, are not noticeable for many years. In the 1960s, they began to reveal themselves in Germany's Black Forest. Scientists found "a baffling form of tree cancer" metastasizing in the famous Schwarzwald.

In ten years, a third of the fir trees in the forest were dead. Half those remaining were nearly dead. "Instead of lush, dark green foliage that gave the Schwarzwald its name, many trees now have only sickly yellowy-brown needles," wrote one researcher. Half the beech trees and half the oaks were also damaged. Elm trees were dying at age 60 rather than their normal life span of 130 years. The freshwaters in Germany's forested mountain regions likewise were succumbing to acidity. Some streams became so toxic the fish disappeared entirely.

Acid rain—more accurately, any acid hanging around the air rain or shine—forms much like the old pea-soupers of London. But instead of smoke we can see, its pollutants are invisible gases we cannot. When industrial smokestacks release sulfur dioxide and nitrogen oxides high into the air, the chemical gases react with the sun, water, and other elements to form tiny drops of sulfuric and nitric acid. These drops, in turn, join up with the clouds. Winds now carry the acids through the atmosphere. Acid rain knows no geopolitical boundaries; it often travels far from the source that created it. One study found high concentrations of the flying pollutants four hundred miles off the Eastern Seaboard over the Atlantic Ocean, presumed to come from U.S. industrial hubs. When it falls back to Earth, acid rain moves through soil, trees and plants, and freshwaters, setting off a cascade of harm.

As the scientists working in the Black Forest puzzled over what was killing trees and lakes, those in the northeastern United States, Canada, Scandinavia, and other parts of Europe also saw the cancer in forests and freshwaters. The sickest areas included the Adirondack Mountains of New York and southern Norway, where hundreds of lakes became fishless. The hot spots all turned out to be upland areas well watered by rain and snow, lying downwind of industrial belts with power stations, smelters, and large cities. Other U.S. hot spots included the Ohio River Valley, the Rockies, the Great Smoky Mountains, parts of Wisconsin and Minnesota, the Pacific Northwest, and the Pine Barrens of New Jersey.

Acidity and alkalinity are measured on what's known as the pH scale, which runs from 0 to 14. Zero is extremely acidic—battery acid. Fourteen is quite alkaline—liquid drain cleaner. Only distilled water has a neutral pH of 7—this is Professor Sabatini's most pristine water, ironically, the one we treat the heck out of.

"Pure" rainwater is already slightly acidic, with a pH around 5.6. Each number represents a tenfold increase, so pH 4 is ten times more acidic than 5, and pH 3 one hundred times more acidic than 5. Most fish cannot survive in waters more acidic than pH 5.

In the 1960s, American scientists sampling rainfall in the White Mountains of New Hampshire were surprised to find rain with a pH of 4 in the quiet Hubbard Brook Experimental Forest, far from any smokestacks. The ecologists F. Herbert Bormann of Yale and Gene Likens of Cornell went on to test rain throughout the Northeast and found pH levels as low as 2.1. In an article in the journal *Science* in 1974, Bormann and Likens connected the deaths of forests and fish in the United States, Canada, and Europe, along with many other ecological impacts they could see or suspect, to acid rain dispersed over long distances by burning fossil fuels.

They sounded the alarm just as Congress was considering loosening the pollution controls of the Clean Air Act of 1970 in response to the era's energy crisis. Instead, the United States and Europe started down a long road to cleaning up the rain. Regulation—especially a cap-and-trade program that lets coal-burning power plants buy and sell emissions permits—has helped reduce the sulfur dioxide released into the atmosphere by more than half. Twenty-five years ago, 18.9 million tons spewed into U.S. skies annually. Today it's well under 8.9 million, the government's cap. In places most prone to acid rain, such as the Ohio River Valley, the pH of rainfall is slowly climbing back from below 4.

While many people who remember the acid rain crisis think of it as "solved," scientists who study acidity as it travels through air and ecosystems say it remains a pressing problem—one with far greater implications than they initially imagined. Acid rain so altered the soils in those sensitive parts of the nation that trees remain more vulnerable

to freeze, disease, pests, and drought. Foresters believe the weakness
has led to extensive die-offs of red spruce trees throughout the eastern
United States, and sugar maples in Pennsylvania. Lakes are healing in
the Adirondacks, but 132 of them remain acidic. Back at the Hubbard
Brook Experimental Forest in New Hampshire, rain has become less
acidic over the past forty years, but it's still ten times more so than
natural rain.

"Yes, we've made significant improvements and the situation is
much safer than it was one hundred years ago and fifty years ago,"
Professor Sabatini told me. "But you cannot look at rain and say it's
pure." That's because the rain is always returning pollution back to us.
Scientists who track contaminants in the atmosphere find that even as
the acid rain picture improves in the United States, rain is spreading
more nitrogen pollution, picked up from the likes of fertilizers and
livestock operations. There is also the reality that while acid rain declines in North America and Europe, it is on the rise in other parts
of the world. In southern China, the stoic Leshan Buddha, chiseled
into cliffs in Sichuan province during the eighth century, is slowly cor-
roding. The largest premodern statue in the world, its face is streaked
with the gray tears of acid rain. At the turn of the century, China surpassed the United States as the largest emitter of sulfur dioxide, and in
2010, India rose to second place behind China. Pollution controls have
helped China begin to reduce its emissions; India's are accelerating
amid rapid economic development, fossil fuel use, and lax regulation.

The atmosphere and its rain are one global system—everything
connected to everything else, sometimes in ways we cannot imagine.
As our understanding of climate shifted from superstition to science,
we learned how an unusually warm sea-surface temperature far out in
the Pacific Ocean, El Niño, can bring torrential rains to California or
failures of the Nile floods in Egypt. How a dust storm formed over the
Sahara Desert can move over the Atlantic Ocean, reducing the odds of
a hurricane slamming into Florida. Now, how humans can change the
rain in strange ways. In South Florida, studies have shown the mas-

sive drainage in the Kissimmee River basin floodplain led to weaker showers over parts of interior Florida. The less water at the surface, the less evaporated to make rain. At the other extreme, researchers now have evidence that the world's largest human-made reservoirs may be spurring extreme rains and flooding in regions such as northern Chile.

Just as Thomas Jefferson suspected, alterations to land and water—clear-cutting a forest, draining a wetland, hoisting a dam, paving a recharge area, large-scale irrigation—all have their bigger-than-butterfly effect on the rain.

At the regional level, agricultural irrigation may be partly to blame for shifting U.S. rainfall patterns, with less falling on the thirsty High Plains region from the Dakotas south to the Texas Panhandle and heavier rains dousing the Midwest. Dr. Jerad Bales, chief water scientist for the U.S. Geological Survey, explains that as farmers pump groundwater up from the High Plains Aquifer and irrigate their crops, the westerlies scoop up the increased water vapor and carry it along, bringing more rain to the Midwest. "As we change one part of the water budget, we change another," Bales says. "It's all linked."

Even cities can change the rain. The heat-island effect is the best-known example of urban impacts to weather; the more paved surfaces and fewer trees in a city, the hotter it becomes. Now several satellite studies conclude that cityscapes influence storms, too. Scientists analyzing precipitation data alongside satellite images of urbanization in China's Pearl River delta found a direct correlation between the rapid growth of cities and a decrease in rainfall. In the United States, a long-term satellite project called the Metropolitan Meteorological Experiment, which analyzes weather patterns over urban areas including Atlanta, Dallas, and St. Louis, found that large cities appear to send rain downwind.

The combination of heat and structures in a city can force up more convective clouds. Skyscrapers can have all sorts of impacts, says the meteorologist Bob Bornstein, who has studied the urban effects of thunderstorms for forty years. They can block sea breezes, act as barriers that force storms around cities, or even split storms in two, tending to send rain to a city's edges.

Now, the warming world has revealed how the emissions we send into the atmosphere—including 36 billion tons of heat-trapping carbon dioxide every year, most from China, followed by the United States—can wreak global climate havoc. A warmer atmosphere and oceans mean more energy aloft that must be dissipated; it finds release in extreme weather. Dry areas get drier, rain falls in torrents rather than gentle downpours. Proof that the strangest rains of all are wrought by humans.

Before Charles Hoy Fort died, his friends, including the writer Theodore Dreiser, founded the Fortean Society; members included the journalist H. L. Mencken and science fiction writers such as Eric Frank Russell. But Fort himself refused to have anything to do with it. He said he didn't believe in the supernatural. He just liked to collect weird facts, use them to satirize authority, and complain how nobody was doing anything about the black rains.

The Forteans endure today, with offshoots around the United States, the United Kingdom, and Australia. A U.K.-based magazine, *Fortean Times,* is still published monthly. Modern writers who've cited Fort's influence include Robert A. Heinlein and Stephen King. In King's novel *Firestarter,* the parents of a child who can start fires telepathically are advised to read Fort's book *Wild Talents* instead of Dr. Spock.

Another fan is the American filmmaker Paul Thomas Anderson. In his 1999 movie *Magnolia,* Anderson directed one of the most shocking rains of film history when he sent grotesque frogs, big as four-pound broilers, pounding from the sky in a heavy rainstorm. At first, it looks as if just a few plump frogs are being flung past the windows of characters involved in the nine-way plot. Then, as the camera holds a shot of a swimming pool, underwater lights glowing in the night, the staggering scale is revealed. Thousands upon thousands of oversized frogs hurl down with the hammering rain—plopping into the pool, thwacking deck and diving board, crashing into trees, and thudding into the street until it's lumpy with carcasses.

Anderson says *Magnolia*'s amphibian apocalypse was inspired by Fort, whose writing on frogs then sent him to read Exodus. *Magnolia* is full of Fortean cameos; the child genius in the film reads *Wild Talents* as he sits in a library cramming to prepare for his appearance on a TV game show. Charles Fort "believed in 'Megonia,' a mythical place above the firmament where stuff would go up to and hang out before dropping back down to Earth. *Magnolia* is a little tribute to that," Anderson told *Variety*. "And it sounds funny, but he believed that you can judge a society by the health of its frogs. That doesn't seem too crazy to me."

It doesn't seem too crazy at all. Scientists say frogs are bioindicators; their well-being reflects the environment's. Because they require both land and aquatic habitat, and have permeable skin that easily absorbs toxins, they are faithful signals of ecological havoc.

There is no greater rain lover than the frog. This truth is held so deeply in India that, when drought takes hold and people are desperate for rain, they'll search out a pair of frogs and organize a frog marriage, replicating all the customs and rituals of a human wedding with hundreds of guests. After the nuptials, the frogs are set free with a prayer for rain.

Frogs symbolize rain for many Native American tribes. Among the Zuni who live beside the Zuni River in New Mexico, frogs are considered the children of six rain priests—U'wanami—who live in houses made of cumulus clouds. The Hopi fashion musical instruments from gourds, a resonator and rasping stick, that mimic the song of frogs to call home rain.

Frogs do seem to call the rain, or at least predict it. Nineteenth-century science journals describe how Europeans kept tree frogs in tall glass jars as wee weather forecasters. Frogs have a "barometric propensity," wrote one chronicler, crawling up their tiny ladders on sunny days, hunkering down in the water below when a storm approached.

In Louisiana, the Creole people say: *Laplie tombé ouaouaron chanté*—

"when the rain is coming, the bullfrogs sing." In my home state of Florida, summer storms are preceded by an ecstatic, escalating ensemble. The weatherwise chorus includes the rain call of the squirrel tree frog—a chattering scold that sounds just like its namesake—and that of the green tree frog, whose rain call southerners claim sounds like "fried bacon, fried bacon."

Many people in the South just call any tree frog a rain frog. As soon as the rain arrives, the mass chorus changes on cue, from rain call to mating call. Frogs often wait for a "stump-floating storm" to breed, explained the late Florida naturalist Archie Carr. Those that lay their eggs in a new rain pool protect their young from the enemies lurking in established ponds, carnivorous water bugs and beetles.

If Fort is right that society shall be judged by the health of its frogs, we're in for a harsh adjudication. Frogs have survived in more or less their current form for the past 250 million years—they made it through the mega-droughts and the pluvials, the ice ages and the asteroids. Today they are vanishing; perhaps this is why frog rains have become so rare in modern times. Nearly two hundred frog species have vanished since 1980 and more than a third of all surviving amphibians are now threatened with extinction, part of a larger catastrophe that scientists call the world's sixth mass extinction.

The rain-loving little bioindicators are definitely trying to tell us something.

THIRTEEN

AND THE FORECAST CALLS
FOR CHANGE

On Memorial Day weekend in 2013, Dave Tzilkowski and his wife, Jillane Hixson, saw dark and massive storm clouds rumbling toward their house in Lamar, Colorado. As they hunkered down, the storm pounded the windows with what sounded like hail, and beat against the door with sixty-mile-an-hour winds. They were trapped inside for fifteen hours.

Tzilkowski and Hixson had been praying for rain, but this storm brought none. It carried a towering wall of dust, as if blowing from the Dirty Thirties. Similar dust storms have rumbled along the flat landscape of southeastern Colorado since 2011, one of the devastating strikes of a drought that climate scientists say is more extreme in this part of the plains than the Dust Bowl that bore down here seventy-five years ago. The past three years have been the driest in recorded history. The tumbleweeds have to be pushed out by snowplow.

The drought and the dust storms have desiccated the wheat crops and the native grasses, too. Prairie normally covered with thousands of acres of cultivated wheat is bare and burnt brown. Farmers' losses are in the millions. Ranchers have sent thousands of head of cattle away to other states. They hope they will not have to relocate their families next.

The same year the Memorial Day duster hit southeast Colorado, in September, copious rains surprised the Front Range on the north side of the state. Residents described the downpours as biblical—or like hell. The storms began on Monday the 9th and did not stop for eight days. Soothing creeks turned to seething rivers. In business districts and neighborhoods, muddy brown floodwaters rose waist-deep. Roads gave way and took cars with them, including a little Subaru carrying four teenagers home from a birthday party in the craggy hills outside Boulder. Floodwaters and mud swept away sweethearts Wesley Quinlan and Wiyanna Nelson, among eight residents who died in the storm. In the largest airlift since Hurricane Katrina, more than a thousand people had to be evacuated from the streaming hills, including an elementary school full of children.

Like the southeastern Colorado drought, the northern Colorado rains broke many previous records, with 9.08 inches falling in twenty-four hours at the peak of the storm. The Boulder area broke every rainfall record on its books—heaviest rain in a day, in a month, and ultimately, in a year. (Not only did the southeastern drought continue after the heavy rains in the north, but more dust storms, strong enough to shut down the highways, rolled in just days later.)

In Colorado, drought and flash floods are nothing new. In July 1976 the Big Thomson Canyon Flood killed 145 people, many of whom were enjoying summer holidays at a popular camping spot an hour northwest of Denver. A thunderstorm lifting over the Front Range stalled and dumped a foot of rain in the canyon in three hours, turning a placid stream into a nineteen-foot wall of raging water.

The 2013 floods, which even the National Weather Service described as out of Scripture, had their meteorological explanation, too. A stationary low-pressure system over the Front Range pulled from two converging plumes of wet weather: a healthy monsoon sweeping out of the southwest from the Pacific Ocean, and a moist tropical system moving north from the Gulf of Mexico. But the question on everyone's minds—from the many atmospheric and climate scientists who live in Boulder, home to the National Center for Atmospheric Research, the National Oceanic and Atmospheric Administration (NOAA), and the University of Colorado, to the thousands of home owners with destroyed houses and no flood insurance—was whether human-caused climate change had made the rains worse.

Scientists can't tie the tragedy of any particular deluge or drought to human-caused climate change, the warming of Earth's average temperature by heat-trapping gasses. We've seen the extent to which our settlement patterns worsen the impacts of natural disasters regardless of what we've done to the atmosphere. But scientists detect the force of climate change in an increasing number of extreme weather events. Human-driven climate change is now impacting all weather, says Martin Hoerling, a climate scientist in NOAA's Earth System Research Laboratory in Boulder. "Rainfall and other weather events happening today are happening in a different climate than one hundred years ago, and so they are behaving differently than they would have a hundred years ago," he says. "The question is how much."

While rain is one of the trickier parts of the atmosphere to measure and to understand, changing rainfall patterns are among the earliest and most obvious tremors of a warming globe. The warmer Earth becomes, the greater the amount of water vapor that forms in the atmosphere. Higher temperatures cause greater evaporation—and therefore more rain—where water exists. They make it hotter and drier where water is scarce. The Intergovernmental Panel on Climate Change stresses that extreme rain events are already increasing in North America and Europe. The scientists predict greater intensity and frequency of both drought and extreme rains around the world as it continues to warm, droughts worsening in arid regions while wet areas become wetter.

The questions of just how much more water vapor will fill the atmosphere in the future—and what the consequences will be—are both unsettled and unsettling. Remember from the early days of Earth that water vapor is a strong greenhouse gas, more potent than carbon dioxide or any other. Water vapor's natural warming has helped keep

Earth in the "Goldilocks spot" that gives us our ideal atmosphere. As the world consumes ever more fossil fuels and temperatures continue to rise, scientists say that the increase of water vapor in the atmosphere could double the greenhouse effect caused by carbon dioxide alone. Some of them, including James Hansen, retired head of NASA's Goddard Institute for Space Studies, believe the hike in water vapor could ultimately cause a "runaway greenhouse." We heard this term used to describe what happened to the rains and oceans of Venus. Remember how our Evening Star was born with water, just like Earth? At some point, Venus was caught in a runaway greenhouse. Its oceans essentially boiled away.

Other climate scientists are wary of the apocalyptic notion, and Hansen acknowledges that it won't happen soon enough for our grandchildren to see it. But they will be around for the storms, evoked in Hansen's 2009 book *Storms of My Grandchildren*. In the years 2011 and 2012, although many parts of the world saw record-breaking heat and drought, the continents collectively drew their two heaviest years of cumulative precipitation since modern record-keeping began.

After thousands of years spent praying for rain or worshipping it; burning witches at the stake to stop rain or sacrificing small children to bring it; mocking rain with irrigated agriculture and cities built in floodplains; even trying to blast rain out of the sky with mortars meant for war, humanity has managed to change the rain.

When I visited the Met, Great Britain's national weather service headquarters, the greatest winter rains seen in British meteorological records dating to 1766 were sopping England and Wales. Founded in 1854 by Vice-Admiral Robert Fitzroy, the brilliant former ship captain who invented sea charts that could "forecast the weather" and later slit his throat amid denouncement of his valiant efforts, the office is located on Fitzroy Road in the historic city of Exeter.

Thanks to the rain obsessions of George James Symons, the Met's scientists have remarkably complete rainfall records dating to the eighteenth century. A cavernous library downstairs holds many of the earliest weather diaries, old ship logs, a weather chart from the D-day landings, the weather records of Robert Falcon Scott's Antarctic expedition, and a sixteenth-century edition of Aristotle's *Meteorologica*.

The climate scientist Mark McCarthy is science manager for the Met's National Climate Information Centre, whose job is to help British government, industry, and citizens better understand weather and climate to manage their risks. As he showed me around the library, he explained that because the nation has been buffeted by violent Atlantic storms for all of its history, and endured so many climatic extremes, such as the storms of King James's and Shakespeare's time, it is difficult to convince people that a warming world is leading to more intense rainfall. That is basic meteorological science. The record-breaking winter of 2013–2014 was yet more telling.

The oldest rain-watching records show that in December through February, the British endured their wettest winter since at least 1766. New station records were set all over England and Wales, with southern England battered by more than double the rain that normally falls in winter. Exceptional gales carrying extreme rains ravaged the southern coastlines, breaching sea walls and turning historic villages into islands. Along the beaches of southwest England, so many severe storms hit the coast that they uncovered hundreds of unexploded shells, bombs, and mines buried on the beaches since World War II. In western Wales, storms stripped away a beach to unearth an ancient forest six thousand years old, said to be the origins of a mythical kingdom, the Welsh Atlantis of folklore and song.

The problem for climate scientists like McCarthy is attribution—figuring out the extent to which climate change drives any particular extreme, from heavy rains to drought—a step they hope will lead to better forecasting of extreme weather. They are getting closer all the time. Scientists from America's NOAA and the United Kingdom's Met brought together researchers from around the world to analyze a dozen extreme events from 2012, including Hurricane Sandy and the drought on the American plains. Increasingly detailed computer climate models crunch thousands of data points, from ocean currents to El Niño patterns, that let scientists tease out the natural workings of the atmosphere from our alteration of it. The study found humans had little impact on the lack of rain in the plains; the climate was much as it had been when Uriah Oblinger migrated west in the nineteenth century. But human-induced warming clearly helped fuel Hurricane Sandy, the largest Atlantic hurricane on record, with winds spanning more than a thousand miles across. Since 1950, sea-level rise caused by climate change has nearly doubled the Northeast's annual probability of a Sandy-scale tempest, the study concluded. In all, six of the twelve extreme events carried the fingerprint of climate change.

When I interviewed McCarthy, he and his team were working on models to predict extreme rains and assess local flood risk on all timescales—hourly for weather alerts, by season for farmers and financial markets, by decade and century to inform planning for infrastructure and climate adaptation. They rely on mega-computers and an exhaustive global network of radars and satellites. And so I was surprised when McCarthy told me about his regular habit of unplugging from his various screens and wandering down to the Met's library. There, he pulls one of the blue hardbound editions of G. J. Symons's *British Rainfall* from the shelf, settles at a table, and loses himself in Symons's obsessive rain recordings and occasional rants. As Symons reported the rainfall at Dublin, Manchester, rain-celebrated Seathwaite, and hundreds of other stations, he also pondered esoteric questions: *What constitutes a rainy day?* Or he'd go on a tear about observers who would set up their rain gauges near a small, unobtrusive tree, then fail to notice over years when it grew up to hamper accurate measurement.

As McCarthy works on the modern mysteries of British rainfall, Symons's century-and-a-half-old rain musings sometimes jar a new idea or spark a new line of inquiry. What tree growing right in front of them might modern climate scientists have failed to notice?

"As well as looking forward," McCarthy tells me, "there is a lot we can learn from the past."

So it is with the history of rain. The links to the past begin with our primitive ancestors, who survived pluvials and droughts to outlive all the other hominids. The prehumans evolved sizable brains, toolmaking smarts, and other survival skills as they adapted to dramatic climate swings in eastern Africa. A big part of what makes us who we are is that flexibility to readjust, revise, and acclimate. It would seem to bode well not only for adapting to the climate we have inadvertently changed, but also for learning to live in ways that don't pull a Venus trick on Earth and its atmosphere.

The best lessons of our rainy past come from the times we made the most of our big brains and tools: when extremes of storm or drought sparked investment and bold new research into meteorological science, launching Fitzroy's forecasts in the U.K. and Abbe's nationwide *Probabilities* in the United States. It is impossible to estimate the millions of lives saved by forecasts and warnings since then.

Today's weather watchers have come around full circle and back to Daniel Defoe, who read so much moral meaning in the great storm of 1703. The journalist Andrew Freedman writes like the best biographers—using the weather's compelling moodiness to interpret the climate's personality for his readers, the 35 million unique visitors to the popular digital news site *Mashable*. As the site's first climate and weather reporter, Freedman covers hurricanes and floods with all the drama of Jim Cantore. But he explains almost every major story about the weather in the context of the changing climate. "Man-made climate change is the biggest weather story ever," Freedman told me. "But I don't want it to be all about the weather. It's not just about a ninety-degree day. Fundamentally this is about sustainability and the global energy choices that we make."

At *Slate*, another of the Internet's new weather nerds, Eric Holthaus, is a meteorologist who built a following by relentlessly tweeting Hurricane Irene and Superstorm Sandy. Holthaus made international news when, after reading the latest sobering report from the IPCC, he announced he would never fly again. The decision was deeply significant for a jet-set science writer with his own pilot's license and a ton of frequent-flyer miles. "My wife and I realized that the 'substantial and sustained reductions' called for by the Intergovernmental Panel on Climate Change had to start with us," Holthaus wrote. "World governments will never agree in time to coordinate reductions in greenhouse gas emissions. If anything is to change, it will have to come from individuals taking ownership of the problem themselves."

Freedman and Holthaus were leading a new generation of weather watchers to not only report the chance of rain and explain the weather, but also educate readers on climate and their role in the future well-being of the atmosphere. They asked their readers to consider the same questions Defoe had asked of his more than three centuries before:

"What can all this be? What is the Matter in the World?"

The science-hostile skeptics of human-caused climate change point to past swings in climate as proof that today's warming is part of a natural cycle. Every major scientific society and 97 percent of the world's climate scientists say otherwise in their consensus that human greenhouse gas emissions are to blame for the current warming. The culture-killing drought of 4,000 years ago has been linked to a centuries-long failure of the Asian monsoon. Causes of other extreme climate events are better understood, such as the volcanic eruptions that marked the Little Ice Age. In contrast, the warming of the past century is not natural in origin; humans have become a dominant force.

To be sure, prior to modern industry and the emissions associated with it, people in the distant past suffered devastating climatic shifts. Given what we've learned about the lost cultures and the tragic times, it hardly seems advisable to plunge headlong into repeating them.

Those who dismiss climate science in the twenty-first century are in danger of repeating the mistakes of the Brits who called ship-forecasting black magic in a time when shipwrecks took the lives of thousands of sailors. Likewise the modern gas and oil companies sabotaging efforts to limit carbon will come to be seen like the vulture ship-salvagers of Cornwall and Devon that claimed the Met's forecasts were putting them out of business.

The history of rain offers hope that our political system is capable of overcoming such selfish interests and our divisions to work together on climate. Congress triumphed when it laid the foundation for the U.S. Weather Bureau to set up national forecasting through the telegraph lines in the wake of the Civil War. The political system worked again in 1990, when Congress amended the Clean Air Act to reduce the pollutants responsible for acid rain. In the quarter century since, the nation's market-based cap-and-trade program has cut sulfur dioxide emissions in half at a fraction of expected costs, one of the most inspirational environmental turnarounds in history. Given its initial championing by Republican administrations and its success combating acid rain, it's ironic to see cap-and-trade now so thoroughly rejected by conservative politicians.

Then there were the times when Congress ignored the advice of its scientists: John Wesley Powell, the first head of the U.S. Geological Survey, who predicted in his *Lands of the Arid Region* report of 1878 how yeoman farmers wouldn't be able to make it on small, unirrigated homesteads in the arid West. Gifford Pinchot, the forestry chief who warned in the early 1920s against the Army Corps of Engineers' "levees only" strategy on the Mississippi River.

In recent years, Congress has resembled the rainmaking 1890s more than the emissions-lowering 1990s—ears open to the influential uninformed rather than its own scientists. U.S. Senator Jim Inhofe of Oklahoma, perhaps the most prominent national opponent of meaningful legislation to reduce fossil-fuel emissions, has said that humans cannot possibly control the climate because only God can do that. "The arrogance of people to think that we, human beings, would be able to change what He is doing in the climate is to me outrageous."

The statement doesn't jibe with Senator Inhofe's vote in favor of the 1990 Clean Air Act amendments that have done so much to combat acid rain. It also ignores fellow Christians who view climate change as one of the great moral challenges of our time.

In addition to his religious convictions, Inhofe has a sense of duty to the energy sector, the largest industry in Oklahoma. The oil and gas industry has donated more than $1.5 million to his political campaigns—more than double the next-largest group of donors. Other evangelical Christians draw a stronger sense of duty from their conviction that humans are called to protect God's creation—and also their neighbors, especially the most vulnerable. "It's the poor and disadvantaged who are being hardest hit," says Katharine Hayhoe, an evangelical and one of the top climate scientists in the United States, "those very people the Bible tells us to care for."

The Great Mississippi Flood of 1927 and the Dust Bowl that followed just three years later were two of many disasters that proved rain does not fall equally on the rich and the poor, the just and the unjust. Neither disaster would have been so devastating for some of the nation's most vulnerable populations had we paid closer attention to the earlier history of river flooding and the drought of Uriah Oblinger's generation.

Whether bringing a river to flood stage or washing fertilizer into the Gulf of Mexico, there's nothing destructive about the rain itself. Only we have made it so. We plowed up the native grasses, settled the floodplains, and built out our cities as if immune to the workings of rain. Today, flooding has become one of the foremost increasing risks of a warming climate.

A growing number of scientists believe we are approaching a climate catastrophe faster than we realize—and suggest it is time to consider large-scale intervention in the atmosphere, known as geoengineering. One scenario has scientists replicating the global cooling that the volcano Mount Pinatubo caused when it erupted in 1991. Ten million tons of sulfur pumped into the atmosphere caused temperatures to drop across the globe by an average 1 degree Fahrenheit. Deliberately pumping thousands of tons of particles into the sky to deflect sunlight would have the same effect, according to the geoengineers who want to research the idea. They liken it to chemotherapy for Earth: Artificial cooling is not something anyone wants to do, but if we have to save the planet, we should be ready with the best possible science. They acknowledge it is worth remembering the drawbacks of chemotherapy, too. The European Geosciences Union has come out with research showing the resulting loss of solar radiation could have troubling effects for people and Earth—including a significant reduction in the rain that falls upon Europe and North America.

As Powell said in his *Arid Lands* report so long ago, "the weather of the globe is a complex whole, each part of which reacts on every other, and each part of which depends on every other." Who could have imagined, in the eighteenth century, a pollutant spewed into the air by Mr. Macintosh's rain-cloak factory in Manchester could help cause a major human health disaster in the Great London Fog of 1952 . . . and then destroy Germany's beautiful Black Forest with invisible and deadly acid rain . . . and then alter the climate of the Earth?

I n the late nineteenth century, a western poet named Joaquin Miller, living in the mountains overlooking Oakland, California, loved rain so much that he created his own personal rain machine to make water roar down on his roof. Anytime he needed some writing inspiration, he could twist a spigot inside his house to summon a shower outside. In 1893, Miller wrote a utopian novel that looked askance at the "extreme selfishness" of a colony of dryland settlers who prayed for rain for their corn even though they knew they would be taking it from their neighbors' thirsty crops of figs. What we now call geoengineering ultimately triumphed in his ideal future society. The impossible caveat was that weather control required utopia, writes the weather historian William B. Meyer—"for no society *short of a perfectly just and harmonious one* was sure to use that power for the better."

In the story of rain and humanity, never entirely just and harmonious, the most shameful injustices took place amid fear and desperation during extreme drought and extreme storms: the sacrifice of children to the rain god Tlaloc, the burning of witches at the stake for conjuring tempests. As Earth throbs with the most extreme rains and droughts of the modern human experience, it is worth remembering those most irrational responses of our stormy past.

In *The Martian Chronicles,* Ray Bradbury wrote that the Martians "blended religion and art and science because, at base, science is no more than an investigation of a miracle we can never explain, and art is an interpretation of that miracle." Whether from the standpoint of science or religion or the art of storm, rain, on balance, has been Earth's blessing. Rain is not only part of our chaotic atmosphere, but part of our chaotic selves—connected in every holy book from the Bible to the Rig Veda, every human genre from cuneiform script to Chopin.

As scientists work to distill its physical mysteries, rain also calls us to breathe its vernacular scents, stamp in its puddles, and cool off in its showers. I couldn't finish the story without celebrating in the rainiest place on Earth, Cherrapunji.

WAITING FOR RAIN

The timing of my arrival in India's state of Meghalaya—from Sanskrit, "abode of clouds"—could not have been more promising. The monsoons, which sweep up from the Indian Ocean on the country's southeast and southwest coasts and fan north to often ecstatic welcome over June and July, normally take until July 15 to cover the nation. But India's Meteorology Office was reporting an early and ferocious season. Monsoons had inundated the entire country as of June 15, the day I arrived in Meghalaya. The last time they had covered all of India this quickly was 1961—more than fifty years before.

So furious were the rains that flooding in the far north had just killed at least eighty pilgrims on their way to the Himalayan shrines near the Tibetan border. Those were the known dead; seventy thousand remained stranded or missing.

Besides the meteorological signs that I was about to experience the rains of my life, another kind of sign sat next to me on my Air India flight to Guwahati, a city in the northeastern Indian state of Assam and the nearest I could fly to reach Earth's rainiest place, Meghalaya's Khasi Hills. A little over halfway through our journey from Delhi, a young Assamese woman with shining black hair that hung to her waist blurted out, "Ma'am, I am not flattering you, but I cannot look away from your beautiful hands, writing and turning pages and highlighting your book."

I could hardly stay buried in my monsoon history after that. So I told her about my rainy mission. With a huge smile, she told me her name: Rimjhim.

Assamese for rain.

Guwahati International is not the smallest airport I have ever seen. But it is the only one with cows wandering around the parking lot among the taxis. My driver, an affable Khasi named Shimborlang Nongrang, who goes by Shim, is here to meet me for a six-hour journey along rutted and sharp-winding roads into Meghalaya and the waterfall-draped Khasi Hills.

Meghalaya is one of India's so-called Seven Sisters, seven states that jut into the far northeast corner of the country, isolated from India's diamond landmass by Bhutan and Bangladesh. Meghalaya happens to be shaped just like a cloud, its southern bottom spread across the northeastern top of Bangladesh. Its village of Cherrapunji, overlooking the Bangladeshi plains, holds the record for rainiest single spot in the world. Between August 1860 and July 1861, the British hill station here measured the greatest rainfall ever recorded, 1,042 inches in one year. More than a third of that came in July on the summer monsoons.

In modern times, Cherrapunji and a neighboring village called Mawsynram vie for rainiest place, both averaging close to 470 inches a year. (Remember, the rainiest metro area in the United States, Mobile, Alabama, pulls in 65 inches annually.) Dozens of the tiny villages nestled in the bright green jungle on this side of the Khasi Hills are probably deserving of the honor. While the monsoons bring their rains to all of India, the intensity here is related to the lift of the winds coming off the Bay of Bengal. Packing the warm monsoons from the Indian Ocean, the clouds riding these winds draw more moisture as they cross the bay, and more heat as they overtake the Bangladeshi plains. As the clouds meet the Khasi Hills and rise over Cherrapunji, Mawsynram, and the neighboring villages perched on the southern slopes, they cool, condense, and drop their record-smashing rains.

I could hardly wait to see cloud-shrouded Cherrapunji. Twenty-five years ago, when the travel writer Alexander Frater wrote his exquisitely rainy narrative of India, *Chasing the Monsoon*, foreigners were prohibited from visiting because of violence at the Indian-Bangladeshi border. Frater wrangled special permission to see the village for a few hours. But the locals were so hostile to outsiders that they wouldn't even sell him an umbrella.

In the quarter century since, the villagers have slowly welcomed eco-tourists who want to enjoy the dozens of falls here, such as Nohkalikai, the highest plunge in India at 1,100 feet, or trek to the area's storied tree-root bridges. Over centuries, assisted by their soaking rains, the Khasis have coaxed the roots of banyan trees across the rivers, creating gnarled, living bridges that look like they belong in a fairy tale.

I was about to experience all of this—amid the heaviest rains in the world. My biggest worry was whether I would be able to see anything in the deluges and cloud cover for which the region is known.

Like many Khasis, Shim speaks the Queen's English, along with Hindi and his native Khasi. "I think this must be your lucky day," he says. It's been an unusually dry June. But now, anvil clouds layered gray like natural stone have gathered to meet us at the Meghalaya border. A fine mist and sweater-worthy drop in temperature hint at rains to come.

'll spend my soggy week in Meghalaya at the Cherrapunjee Holiday Resort, a decision based on the exuberant appreciation for rain and the monsoons expressed on its website by the family that owns the place. The resort is twenty kilometers beyond Cherrapunji; the village proper is barren, its forest clear-cut and the limestone beneath its ground constantly gouged—rock trucks come and go like ants to a picnic.

Denis Rayen and his wife, Carmela Shati, built their resort amid the well-watered jungle and tiny villages halfway between the rain competitors Cherrapunji and Mawsynram. Shati is a native Khasi who met Rayen when he visited Meghalaya as a field officer evaluating projects for an NGO. After they married, he became a successful banker, but he was ill-suited to the profession. As they followed his career around India, he fantasized about returning to Shati's native village someday to open a hotel catering to monsoon tourism. "Everywhere we traveled, when people found out she was from Cherrapunji, they knew it as the rainiest place on Earth," Rayen says. "Then, they would want to know if they could visit. But it wasn't easy to visit."

During dry bank meetings, Rayen would doodle sketches of his wet resort, a place where families could venture out together on treks, and where he could market something unique in the world. "I didn't want to sell hotel nights and I didn't want to sell food," Rayen tells me. "I wanted to sell rain." He and Shati started looking for land.

The Khasis are matrilineal, meaning the women own the land and pass it on to their daughters. (Not as good a deal as it sounds: The daughters who inherit the family's property also inherit its misery— bearing full responsibility for aging parents; siblings who lose their jobs or spouses; down-on-their-luck in-laws; alcoholic or drug-addicted relatives; and other familial burdens.) In 1998, Rayen and Shati found their dream spot atop a steep hill in the village of Laitkynsew, with a view of the Khasi Hills and their plunging waterfalls to the north, the Bangladeshi plains to the south. The village doesn't allow outsiders to own property, but as a native, Shati could buy it in her name. She stayed in Shillong with the couple's small son and daughter while Rayen went to work building their resort in the rain.

"Resort" is a relative term. My room is sparse, and unscreened windows mean sharing space with various jungle bugs and the lizards hunting them. The view, though, is rich. Awake at dawn the first morning, I step outside into a cauldron of clouds. It looks as if the entire sky has dropped to eye level. A white-gray mist swirls before me, and in the valley below. Only the peaks of the Khasi Hills are visible in the brew.

By 8 a.m., the swirls begin to disappear. Bit by bit, they reveal a blue sky that soon takes over the day.

---

am relieved to have one cloudless day in the rainiest place on Earth. To see rain's colorful legacy requires a little time without. The sunshine also saves me some money. I'd planned to hire guides to trek through the Khasi Hills in the monsoon. But today's visibility is such that I'll be able to navigate the village and jungle footpaths with only a map.

As I begin climbing down to the valley of the tree-root bridges from a village called Tyrna, I am reminded once again of the unfairness of rain's gray reputation. The rainiest place on Earth throws off the greenest horizon I've ever seen. During the Raj, the homesick British dubbed the forested hill region "Scotland of the East."

Closer up, rain's brush has trimmed the green jungle and villages in vivid primary colors. The villagers, who have erected their tiny houses on the slopes with a hodgepodge of materials, devote more square footage to flowers than floor space. Yellow trumpet flowers, which bloom on delicate vines in my part of the world, grow here into gnarled trunks that explode their sunny blossoms like fireworks. Roses thrive in the small yards under skinny coconut palms and thick jackfruit trees. As the village of Tyrna disappears behind me and the jungle opens ahead, the flowers increase in number, chaotic not cultivated, run amok as if drunk on rain. The magnified monsoons make this subtropical forest one of the wettest eco-regions in the world, and one of the richest botanical habitats in Asia. Scientists have identified 250 orchid species in this part of the Khasi Hills. Many of them bloom in June. The delicate ground orchids are easiest to see, voluptuous pink and purple petals calling attention from the forest floor. Clinging to trees, the larger epiphytes are harder to spot in the orgy of supersized leaves, shoots, bulbs, aroids, pipers, ferns, berries, nuts, and fruits familiar and strange. I spot figs as big as my fist and jackfruit larger than my head.

The jungle bounty sustains Khasi culture and economy. The villagers have traditionally made their living cutting firewood, making charcoal, and trading an array of forest goods: bamboo shoots, wild vegetables, honey, mushrooms, areca, betel, other nuts, and a high-piled tropical fruit bowl. But the real nourishment for both forest and people is the rainwater that pours from the clouds, hangs in the air, dives from the waterfalls, rushes to the rivers, flows into clear pools, and fills the streams and rivulets that run through the villages, tapped by each family with its own Rube Goldberg arrangement of bamboo pipes and handmade filters.

The rain also keeps the jungle and the villages cool; historical mean temperature for June, the warmest month, is 68.5 degrees Fahrenheit. This afternoon it is 80. That wouldn't be so bad but for the humidity— and the fact that the stone footpath has turned to what feels like the longest staircase in the world. Two thousand narrow steps plunge before me, a direct drop into the valley. I'd trained for these stairs by hiking in hot, humid temperatures up and down my city's anemic hills. It is not preparation enough for a sun-scorched afternoon in Cherrapunji. Halfway down the stairs, my shirt and backpack are so soaked I'm sure my water bottle has spilled. But it is only sweat.

The tree-root bridge I'm looking for spans the Simtung River. I'm so hot by the time I find the bank that I jump onto a flat boulder and plunge my head into a waterfall-fed pool before I even see the bioengineered wonder. It is up the river, behind me. The living footbridge and its ropy handrails stretch ninety-five feet from bank to bank, handiwork of nature and people entwined. It's like *The Swiss Family Robinson* come to life; I imagine the novel's littlest boy, Franz, might race across at any moment.

The second morning that I awaken at the holiday resort, I learn that I am not the only journalist hanging around Cherrapunji, India, waiting for rain. An entire crew of TV weather personalities, storm chasers, and photographers from the BBC has arrived at the resort for their work on the children's program *Fierce Earth*. Having produced episodes on extreme cold, extreme hot, tornadoes, wildfires, floods, earthquakes, hail, thunderstorms, tsunamis, volcanoes, and hurricanes, the team has come to Cherrapunji during the rainy season to film *Fierce Earth: Monsoons,* scheduled for the fall.

By the looks of the sky, the only extreme footage likely today is for *Fierce Earth: Sunburn.*

The resort's rooms open to an octagon-shaped dining area where the guests sit together for meals and talk incessantly about the clouds, the weather forecast, and especially the prospects for rain. I wince to hear one of the storm chasers, an American, reveal that he doesn't think humans are responsible for climate change. While the scientist in the group, British geologist Dougal Jerram, tries to set him straight, I wander off to find Denis Rayen.

Rayen is at his computer screen, frowning at satellite images of cloud cover moving across southern Meghalaya. All month, the clouds have swept up from the Bay of Bengal as usual. But they seem to move on without bringing the rain. In an average year, Cherrapunji's June rainfall is nearly 100 inches. June is half over, and only 23 inches have fallen.

"I'm afraid it's not going to rain again today," Rayen tells me in his soft lilt. He seems almost too civilized to live in the jungle. Greeting his guests in the dress shirts, pressed pants, and leather shoes of a banker, he is not the type of resort owner to hang out in a pair of flip-flops.

And he is perhaps the only one in the world so deeply troubled by the prospect of another sunny day in paradise.

I spend the morning at the small wooden desk I've pushed up against the window in my room, in turn writing and scouring the horizon for a promising cloud. The sky is pale blue, the field of vision so clear that I can see across the Bangladeshi plains to my left, the Khasi Hills to my right, and the village of Laitkynsew ahead, wee houses stacked on a hill in the distance.

For the afternoon, I plan to hike in the hills around Laitkynsew,

Sosarat, and Siej villages to meet more of the people who live in the rainiest place on Earth. Besides my water bottles and camera, I tuck my raincoat and rainpants wishfully into my backpack—along with rainproof notebook and all-weather pen.

I set off in step with a piercing chorus of cicadas. They remind me of the dullest and most sweltering summer days of my childhood. Florida's cicadas hum the loudest on the hottest, sunniest, and stillest afternoons. Now their Cherrapunji cousins are buzzing more intently with each step I take, as if to rub in the sunshine. It is not a sound I expected here; cicadas don't like singing in the rain.

The village circuit is hot and steep. But the people I pass in the road are refreshing like breezes, especially the children. Many of the women and girls hold umbrellas to shield themselves from the sun. Several are also hauling buckets and pots to the community water taps. As I saw on yesterday's trek, most of the homes have water pipes rigged from small streams, pools, or waterfalls. The rainiest place on Earth is prone to increasingly severe water shortages during dry times.

Soon my clothes and backpack are wringing wet again. Trudging up a steep incline in Siej, I meet a villager walking down, a pretty older woman with a thick black braid and a yellow umbrella. She asks me where I'm from. I tell her, and explain my trip to Meghalaya.

"I hope it rains," I say.

"Oh! It's going to rain!" she exclaims, pointing to the southern skies. The clouds still look gauzy. But they've dropped lower, and they seem to be dragging some friends behind.

"When?" I ask her.

"It's going to rain last night!"

If not the grammar, the cloud-watching villager is right about the rain. At 1 a.m., I am awakened from a deep sleep by the sudden roar of rain on rooftop. It is as if every god worshipped in India (there are said to be millions) has turned on their heavenly spigots at once—all fullblast atop the head of little Cherrapunji. I jump from bed and pull my curtain to capture the moment my weather fortunes have changed. A Pluvius-worthy lightning strike spotlights the rain like the rock star it has become for those of us waiting at the resort.

I hurry outside to stand under the front awning with other bare-footed guests—a Mumbai petroleum engineer and his teenaged son, and part of the BBC crew. As drops bounce our way from puddles, I think of the orchids pollinated by rain splash, an evolutionary strategy known as *ombrophily*. The spattering feels cool, William Carlos Williams's "certain unquenchable exaltation."

Within a disappointingly few minutes, though, the holy torrents lessen to mortal rainstorm. In twenty minutes, it has all moved on, rain, thunder, and lightning receding into the Khasi Hills like a dream.

We learn in the morning that it was not a dream; old Jupiter's lightning strike has knocked out some of Rayen's transformers, including those responsible for the resort's wi-fi, and the power in my room. But the day has dawned too beautifully—and too sunny—to stay indoors. I decide to head into Cherrapunji for market day, held once every eight days, rain or shine.

In the octagonal dining room, the BBC team is mapping out a day trip to film at some of the region's signature waterfalls. This is Plan B. The waterfalls will give them the sights and sounds of rushing water in the rainiest place on Earth in case of the unthinkable—a sunny June week in Cherrapunji.

Carmela Shati graciously lets me tag along on her regular trip to the Cherrapunji market. From a core block of permanent stalls, the market spreads into the sidewalks and streets of the city center in a colorful, crowded patchwork of vendors, most of them women sitting on plastic sheets and surrounded by artful heaps of produce. Jackfruit are piled like boulders. Tiny green and red peppers spill over from baskets. Familiar veggies have lyrical names—okra is *lady's finger*, eggplant *brinjal*, bay leaf *tezpatta*. The rainbow of mangoes alone could fill a Western grocery's produce section. Other corners of the market are devoted to meats, and to local crafts, which include the Khasis' famous handmade locks, knives, and bows and arrows.

Shati attended school here in Cherrapunji, five kilometers from home. This time of year, it rained so intensely and steadily that she and the other children could not walk home for weeks at a time. They bunked in the school's hostel, waiting for a clear day to return to their families. "It was common to not see the sun for ten days or more," Shati tells me.

Such rains have been scarce since she and Rayen returned to Cherrapunji. "It used to rain steady, for days. If it rains now, it will be fierce all at once, and then this," she says, pointing to the bright sky. The changed rain has changed the feel of summer. In the past, no one owned a fan, or opened umbrellas against the scorching sun. Both are necessities now.

Once Shati's bag is filled with cabbages, bulging peas, and bulbs of ginger for the resort's kitchen, we take a walk to visit her friends, an elderly couple called the Elayaths who live atop a hill overlooking Cherrapunji. Even on this sunny day, their neat white house is surrounded by clouds. Raman Elayath is a retired beekeeping expert who came to Cherrapunji in 1964 to help jump-start the local honey industry. That year, he says, it rained 48 inches in a single day. In his first decades here, he never slept under a fan in summer like he did last night. He never went around in a sleeveless T-shirt like the one he's wearing now.

Like every other local I've met, the Elayaths have no doubt the climate is changing, and that humans are changing it. They are mystified that anyone anywhere else in the world thinks otherwise. "For us, it is not changing slowly," Raman Elayath says. "It is changing rapidly." Besides global warming that may be shifting Cherrapunji's signature rains, Elayath has watched rampant water diversions, mining, and deforestation damage the landscape and the livelihood of his adopted home.

On the drive back to the resort, I enjoy the cheesy yellow tourism signs that Rayen has posted at teasing intervals alongside the steep and curvy road:

"Experience the monsoon."

"Romance in the rain."

I make up one of my own: Hike in the heat wave.

On the fourth morning I wake at 5 a.m.—to rain, but not nearly as much as my family is getting back home. Had I never left, I would have gotten to see the greatest Florida rains of my lifetime. Green moss has begun to grow on the side of our house, our rain-hating hound won't step outside, and our car has developed a bizarre leak that sends rain into pools on the driver's-side floorboard. Florida is drowning in the rainiest summer in state history while I lament the paltry precipitation in the rainiest place on Earth.

The BBC crew is up too, frantically readying equipment to at last document the rain in the rainiest place. While the technicians grab up video cameras and umbrellas, raincoat-clad weather personalities Clare Nasir and Mike Theiss run around outside to soak themselves in a storm that is surely lighter than she is used to in her rain-famous home of Manchester, or he in the even-rainier Florida Keys. Duly dripping, they practice lines out front of the resort, where the million-dollar view is now one giant cloud of morning fog:

"Battered by rain and storm clouds, THIS is the wettest place on Earth!"

"It's not just raining cats and dogs. It's raining elephants and rhinos!"

A cameraman starts filming. I hear Nasir tell her young audience a white lie—or maybe it's a gray one: "It's been raining here for days!"

Like yesterday's, this rain doesn't last the morning. But I've made more certain arrangements. I've e-mailed rain an invitation to accompany me on this afternoon's trek.

Manoj Gogoi was seven years old when his mom, pregnant with his little sister, went into labor three months prematurely. It happened during a heavy thunderstorm. Manoj was the one who called for help. While his mom was rushed off to the hospital, he stayed back at the family's apartment in Guwahati. During the hours he spent praying for his mother and baby sister, he watched the rain sheet outside his window. When he learned they were both okay, his little-boy brain conflated their survival with the storm. He insisted his parents name the new baby for the rain—Rimjhim.

Rimjhim Gogoi is now twenty-one and in glowing good health, an undergraduate student in computer-science engineering in New Delhi. My new friend from the plane, she has come home to Guwahati, in the state of Assam, to spend summer break with her parents.

Like Meghalaya next door, Assam has a tropical monsoon rain forest climate, with cool summer temperatures and lots of rain. But this month has smashed all heat records. The week we flew in, Guwahati and Assam hit their highest temperatures in history, 38.8 degrees Celsius in Guwahati, or near 102 degrees Fahrenheit. Across Assam, twenty-six people died of heatstroke and related illnesses in two days. Local governments shuttered the public schools, which don't have air-conditioning; they generally don't need it.

Thinking of the six-hour drive from Guwahati along hairy roads, I did not imagine Rimjhim would take me up on my invite to go trekking. But right on time, a tiny taxi crunches up the steep, gravel road and Rimjhim jumps out—followed by her mom. Dressed in strappy sandals, Mom insists she'll be happy spending six more hours waiting for us. I hand her the key to my room, and Rimjhim and I set off for the most spectacular tree-root bridge, known as the Nongriat Village Double-Decker.

The morning's clouds have cleared the sky like an eraser, leaving a washed-out slate. We head for the two thousand stairs. After the Swiss Family Robinson bridge, we'll hike again that far, descending deeper into the valley, climbing up part of the other side, and crossing two rivers on suspension bridges made of steel-wire rope.

A small, white sun burns through the white atmosphere, making the two thousand steps feel like two thousand oven vents. But they go

more easily with someone to talk to. Down the first thousand, Rimjhim and I cover our families, and love marriages versus arranged.
During the second, as we begin to drip, Rimjhim has a confession. She
despises sweat. This does not surprise me. I had noted, without judgment, I hope, the chichi dress on the airplane; the red-manicured fin-
gernails; and the e-mail address that contains the word "fashionista."

Her next string of confessions is more substantial. She delivers
them at the first suspension bridge, which stretches some thirty feet
above the Simtung River. Rimjhim is terrified of heights. And water.
And she can't swim.

I am flabbergasted. True, I never mentioned the steel footbridges.
But the words "tree-root bridges" and "rainiest place on Earth" would
have dissuaded most nonswimming acrophobics from this trek.

"I am so sorry, ma'am," she tells me. "I'm afraid I cannot cross."

The bridge's base does seem meager—just eight steel ropes, an inch
of sky between each. Wire woven between the base and steel-rope
handrails, presumably for safety in the monsoons, is rusted from all
the rain, and springing out here and there. Since it hasn't been raining
as much as usual, the Simtung isn't raging beneath us, or swollen as
it normally would be in June. Through the clear-rushing water, I can
see the boulders that line the riverbed. If the bridge gives way, I don't
think Rimjhim will drown. She'll just be crushed on the supersized
rocks.

Having already e-mailed with her brother to learn the story behind
her name, I know that Rimjhim is the beloved miracle baby of her
family, center of the universe to her brother, father, and sweet mom
waiting back in my room. Still, after 8,000 miles and 2,000 steps, I am
not inclined to turn back before I get to see the Double-Decker Root
Bridge grown strong in the rain.

I do the only thing I can think of, which is to tell some serious lies.
Lies worse than the white or gray ones Clare Nasir told the children
watching *Fierce Earth: Monsoons*. Lies worse than any I've told my own
kids to convince them okra is *not that slimy*, or that the end of the

hiking trail is *just around the corner*. I tell Rimjhim *I am absolutely certain* this bridge is rock-solid, that it is safe, that we will make it across fine, and that all she has to do is hold on, put one foot in front of the other, and follow me to the other side. "Just don't look down."

I take Rimjhim's gear, step onto the bridge with feigned confidence, and tell her we will be across in no time. This is true for me; my fear makes me scramble, water shoes slipping on the ropes. Rimjhim takes the bridge impossibly slowly, gripping the handrails and looking down with every step. When her feet touch soil on the other side, she is drenched in the dreaded sweat. She is shaking. But her enormous smile is back. She is thrilled to have conquered the bridge, and we snap triumphant pictures.

The next suspension bridge is higher, longer, and flimsier. It sags where the other was taut. As an added bonus, a sleepy-looking villager is squatting near our end, making a repair. He signals us to wait, and we go through an almost identical sequence: Rimjhim eyes the barefooted repairman and says she can't cross. I tell her lies as big as the boulders underneath us.

The water in this river, the Umkynsan, is faster-moving and deeper, rushing down the boulders in rapids and flowing into gorgeous turquoise pools. When the villager waves us across, I step onto the bridge and try to ignore the bamboo and other mysterious, shifting patches covering gaps in the steel-wire rope. Tortoise-like, Rimjhim follows me across. More triumph, more pictures, more sweat.

After the suspension bridges, our final climb up moss-grown stairs and stones into Nongriat Village feels easy. Gateway to the Double-Decker, Nongriat announces itself with a small, handwritten sign and a trash can made from a mustard-oil tin, both nailed to a tree. A tight bouquet of yellow butterflies, clinging lower on the tree trunk, blossoms out into flight as we walk by.

Through the tiny village and down one more flight of mossy steps, we finally lay eyes on the monument built by generations of villagers, roots, and rain.

Seeing the pair of bridges is divine, like spotting a double rainbow: One alone is already a masterpiece. The double-decker tree root bridges seem more a miracle. They stretch across a gorge from a giant *Ficus elastica*, an Indian rubber tree that has grown over and around large boulders to become part of the riverbank. Untold generations ago, villagers began threading thin new roots across to the opposite side. They used long, hollow betel-nut branches and flat stones to create the base, and molded handrails, too. Now the living roots have grown huge, and hardened into two vine-muscled walkways with waist-high walls, all sturdy as concrete. (As if anyone will trust me on the topic of bridge sturdiness.) Most constructed bridges, from the two we crossed this afternoon to the urban freeways of Delhi or Detroit, weaken over time. The double-decker has grown stronger.

I am deeply disappointed that I have missed the monsoons. I am also mad at myself for the magical thinking that Rimjhim would bring rain—and a tidy ending to the story. How easily I succumbed to my own version of praying for rain, no different than the governor of Georgia clasping hands at the gold-domed capitol in Atlanta or the Sumerians worshipping Iškur, the lightning-bolt-wielding god standing on the back of a bull.

Rayen and Shati fear that Cherrapunji's rainfall has changed permanently since Shati's girlhood; that Meghalaya will lose its "Rainiest Place on Earth" title to global warming. Scientists say this is not yet possible to conclude. The latest climate models suggest changing patterns to the Indian monsoon will mean more extreme dry spells, but also far more intense rainfall in the future as the oceans warm. By the time I left India, the extreme rains that had surprised religious pilgrims at the Tibetan border in the north had killed more than five thousand people in flash floods and massive landslides. While Cherrapunji pined for rain, entire villages and towns in northern India had been washed away.

When I finish out my last two days in Meghalaya, both will measure zero on the rain gauge, a rarity for June even in these past few years of rainfall deficits. But admiring the double-decker root bridge at Nongriat, I see rain in every pixel: Rain has filled the pools where a villager is washing her clothes and Rimjhim is having a splash. The young woman afraid of water is enjoying it—the need to drink, to cool off, and to celebrate a journey's end now greater than her fear.

As it has quenched humanity, rain has nourished this jungle, carved this gorge, fed these waterfalls, and made this stream, the Umshiang, gentle today but raging during the monsoons, tempests that led someone long ago to dream of protective treeways in the sky. The storybook *Ficus elastica* is part of the banyan family, and rain is the banyan's mother's milk—even the reason for its heart-shaped leaves. The "drip-tips" guide rainfall gently to the ground, protecting soil from the pounding monsoons. Roots dripping from its canopy like showers, the mythical banyan tree needs no soil to get established, no groundwater to drink. The water and nutrients that give it life come entirely from rain.

ACKNOWLEDGMENTS

Of the many people who helped with these pages—and helped care for my two exuberant children so that I could write these pages—no one has contributed more than Aaron Hoover. As well as my husband, Aaron is my first editor, a speechwriter who elevates my words as he does my life and our children's.

Will and Ilana Hoover inspire everything I do. I thank them for being remarkably good sports—patient with my work and the sort of kids who not only swim, hike, and bike in the rain, but always aim for the puddles.

I am grateful for my writing partner, environmental historian Jack Davis, whose wisdom, critiques, and support helped shape *Rain* from the day I had the idea through the weeks, months, and years I worked to turn it into this book. *Rain* is also much better thanks to three writers who became trusted friends over our shared interest in water and words, Heather Dewar, Emily Green, and Christine Klein.

At Crown Publishing Group, I thank executive editor Rachel Klayman and publisher Molly Stern for believing in a book of rain; Domenica Alioto for her kindness and skillful editing; and Emma Berry for a million details. I treasure Anna Kochman and Chris Brand's design, classic as an umbrella, and I thank production editor Christine Tanigawa and copy editor John McGhee for their saves. Dyana Messina and Danielle Crabtree are tireless promoters like the old-time rainmakers. At the Sandra Dijkstra Literary Agency, my thanks to Elise Capron, known in our house as my secret agent.

I owe debts of gratitude to umpteen meteorologists and other scientists, none more than Greg Hammer at NOAA's National Climatic Data Center in Asheville, North Carolina, who was there from the time I set up initial interviews to the final days of fact-checking. In addition to the atmospheric and other scientists quoted in these pages, I am obliged to many others behind the scenes, all devoted to helping the public understand weather and climate and their risks. They include Dave Easterling and Mike Brewer at NCDC and Kevin Kelleher and Jonathan Gourley at the National Severe Storms Laboratory in Norman, Oklahoma. I thank University of Oklahoma professors Bob Puls and David Sabatini, and Renee McPherson with the Oklahoma Climatological Survey. At the Met in Exeter, England, I thank Dan Williams and ever-patient Mark Beswick.

Special thanks to Melissa Griffin, assistant state climatologist of Florida, for the many years she has answered my questions on the vagaries of weather and climate.

At the University of Florida, I am indebted to the librarians, especially Florence Turcotte, and to scholars including Glenn Acomb in landscape architecture, Pierce Jones in extension, Joe Delfino in water resources, Paul Mueller in geology, and Bron Taylor in religion. For her warm and relevant introduction to India and Hinduism I am grateful to the Hindu scholar Vasudha Narayanan. At the University of Utah, thanks to Robin Craig and Robert Keiter.

I thank *Orion* magazine and managing editor Andrew Blechman for the opportunity to write about water in Seattle, work that made its way into several of these chapters. The Seattle rain artist Buster Simpson is a font of information and inspiration.

In Scotland, I thank Willie Ross and Daniel Dunko, and in England, Cambridge University Museum of Zoology Director Paul Brakefield. In India, a river of thanks to Jose Kalathil, Shakti Vinay Shukla, Dr. Ramesh Srivastava, and Brijesh Caturvedi; to the Rayen family: Denis Rayen and Carmela Shati and their children, Angela and Joel; and to the Gogoi family, rain-named Rimjhim, Manoj, and their parents, Dipti and Dilip.

The Society of Environmental Journalists and its members have been incomparable sources of support. Many SEJ members, including William Souder, Robert McClure, Bill Kovarik, and Craig Pittman, came through when *Rain* needed them. I am indebted to other SEJ members including John Fleck and Nancy Gaarder for inspired reporting on climate change, weather, and rain.

I owe much to the Escape to Create artist residency, Marsha Dowler, Karen Holland, my fellow E2C artists, and the community of Seaside, Florida—a town that proves the influence of urban design (and rainstorms) on creativity. I wrote my favorite chapters at the cottage of Seaside founders Robert and Daryl Davis, watching storms move in from the Gulf of Mexico. I am grateful to Jane Toby for caring for the children then. For other cherished writing spaces, I thank Bill and Julie Pine and the island of Cedar Key, Florida; and Bob Knight and Debra Segal and the community of Celo, North Carolina.

My heartfelt thanks for the specific assistance of Anthony Anella, Karen Arnold, Maribel Balbin, Kate Barnes, Laura Bialeck, Mindy Blum, Joe Browder, Julie Brown, Michael Campana, Gracy and Mike Castine, Brenda Chalfin, Ronnie Cochran, Mary Furman, Lesley Gamble, Gerry and Joe Garrison, Kim Gregg and the Rev. Steve Gregg, Thomas Hallock, Melanie Hobson and Charlie Hailey, Paul Hoover, Yasuko Horie, Larry Leshan, Jacki Levine, Dr. Bernie and Chris Machen, Whitey Markle, Karl Meyer, Emily Monda-Poe, John Moran, Pat Morse, Louise OFarrell, Melissa Orth, Jim and Claude Owens, Annie Pais, Nancy Peck, Jonathan Rabb, Walker Roberts, Sonya Rudenstine, Ros Sadlier, Steve Seibert, Dan Smith, Kristen and Dan Stoner, Makiko and Dave Waldrop, Virginia Walsh, and Clive Wynne.

I owe special thanks to Stephen Mulkey, president of Unity College in Maine.

Finally, for a book this wide-ranging, I sought help from many more experts than I am able to acknowledge here, including the authors in the source material. I thank them all and absolve them of any responsibility for my interpretations and any mistakes, mine alone.

NOTES

PROLOGUE: ORIGINS

1 **Scientists viewed Mars:** David Harry Grinspoon, *Venus Revealed: A New Look Below the Clouds of Our Mysterious Twin Planet* (New York: Perseus Publishing, 1997), 49–51.

1 **Hawking newspapers:** Sam Weller, *The Bradbury Chronicles: The Life of Ray Bradbury* (New York: Harper Perennial, 2006), 43. For Bradbury's love of rain, see p. 90.

2 **Earth, Mars, and Venus were born:** Author interview with Dr. David Harry Grinspoon, NASA/Library of Congress Chair in Astrobiology, April 9, 2013.

3 **This is the beauty:** C. Donald Ahrens, *Essentials of Meteorology: An Invitation to the Atmosphere* (Belmont, Calif.: Brooks/Cole, Cengage Learning, 2012). Professor Ahrens's *Essentials of Meteorology* and *Meteorology Today: An Introduction to Weather, Climate, and the Environment* (Belmont, Calif.: Brooks/Cole, Cengage Learning, 2012) are indeed essential for anyone who wants to understand weather and climate. For a primer on water vapor, see *Essentials*, pp. 5, 84–87.

4 **Nature's trustiest timepiece:** Author interview with Dr. Paul A. Mueller, University of Florida Department of Geological Sciences, Gainesville, Florida, February 27, 2013.

5 **The irony:** Grinspoon, *Venus Revealed*, p. 31.

6 **Mars, too, appears:** Author interview with Grinspoon; and David H. Grinspoon, "Chasing the Lost Oceans of Venus," in Michael Carroll and Rosaly Lopes, eds., *Alien Seas: Oceans in Space* (New York: Springer Publishing, 2013), 3–10.

6 **Venus cooked:** Author interview with Grinspoon; and Grinspoon, "Chasing the Lost Oceans of Venus," 3–10.

7 **"Sunshine abounds everywhere":** John Burroughs, "Is It Going to Rain?" *Scribner's Monthly Illustrated Magazine*, July 1878, 399.

8 **"The earth has enough":** Thomas Jefferson to James Madison, May 23, 1806, Monticello. Thomas Jefferson Papers, Library of Congress.

11 **Amid the worst drought:** Hadley and Peter Arnold, "Pivot: Reconceiving Water Scarcity as Design Opportunity," *BOOM: The Journal of California,* vol. 3, no. 3 (January 2013), 97.

11 **Globally, the continents recently drew:** NOAA National Climatic Data Center, Global Analysis, 2011. The 2011 globally averaged precipitation over land was the second-wettest year on record, behind 2010, with greatly varied rainfall across the world including severe drought in the Horn of Africa.

ONE: CLOUDY WITH A CHANCE OF CIVILIZATION

16 **The rainy-day butterfly:** Author interview with Paul M. Brakefield, November 25, 2013; and Paul M. Brakefield, Julie Gates, Dave Keys, Fanja Kesbeke, Pieter J. Wijingaarden, Antonia Monteiro, Vernon French, and Sean B. Carroll, "Development, Plasticity and Evolution of Butterfly Eyespot Patterns," *Nature,* vol. 384 (November 1996), 236–42.

16 **As the drops hit:** Xu-Li Fan, Spencer C. H. Barrett, Hua Lin, Ling-Ling Chen, Xiang Zhou, and Jian-Yun Gao, "Rain Pollination Provides Reproductive Assurance in a Deceptive Orchid," *Annals of Botany,* vol. 110, no. 5 (May 2012), 953–58.

16 **An exhaustive study:** Brad T. Gomez, Thomas G. Hansford, and George A. Krause, "The Republicans Should Pray for Rain: Weather, Turnout, and Voting in U.S. Presidential Elections," *Journal of Politics,* vol. 39, no. 3 (August 2007), 649–63.

16 ***Les Misérables*:** Laura Lee, *Blame It on the Rain: How the Weather Has Changed History* (New York: Harper, 2006), 166–69.

17 **"Providence needed only":** Victor Hugo, *Les Misérables* (New York: Dodd, Mead and Co., 1862), 112.

17 **Once prospering in vast numbers from the riches of the rain:** Timothy Egan, *The Good Rain: Across Time and Terrain in the Pacific Northwest* (New York: Vintage, 1990), 36.

18 **An anti-noise-pollution organization:** "One Square Inch," a sanctuary for silence at Olympic National Park, http://onesquareinch.org/.

18 **Connecting East and West:** "The tide-beating heart of earth" is from Melville's memorable description of the Pacific in the chapter of the same name in *Moby Dick.* Herman Melville, *Moby Dick, or, The Whale* (Originally published London: Richard Bentley, 1851. Quote is from Los Angeles: Arion Press, 1979), 491.

19 **As the heavy ocean plate:** Eugene P. Kiver and David V. Harris, *Geology of U.S. Parklands* (New York: John Wiley & Sons, 5th edition, 1999), 77–78; and "Olympic National Park," National Geographic Travel, http://travel.nationalgeographic.com/travel/national-parks/olympic-national-park/.

20 **After this warm, wet air:** Ahrens, *Meteorology Today,* 507.

20 **In 1860:** Ibid., 350.

20 **A fast-growing haven:** Author interview with Dr. Clifford F. Mass, University of Washington Department of Atmospheric Sciences, Seattle, February 1, 2013; and Cliff Mass, *The Weather of the Pacific Northwest* (Seattle: University of Washington Press, 2008), 18.

21 **They found the circulation:** Clifford Mass and Carl Sagan, "A Numerical Circulation Model with Topography for the Martian Southern Hemisphere," *Journal of Atmospheric Science*, vol. 33 (1976), 1418–30.

21 **This rain shadow covers:** Author interview with Mass; and Mass, *The Weather of the Pacific Northwest*, 19–20.

22 **The number of days:** Author interview with Mass; and Mass, *The Weather of the Pacific Northwest*, 20.

22 **While Seattle experiences:** Christopher C. Burt, "Thunderstorms: The 'Stormiest' Places in the U.S.A. and the World," Weather Underground, June 21, 2012, http://www.wunderground.com/blog/weatherhistorian/thunderstorms-the-stormiest-places-in-the-usa-and-the-world.

23 **New Orleans and West Palm Beach:** Data courtesy Scott E. Stephens, meteorologist, NOAA National Climatic Data Center, Asheville, North Carolina. The top ten cities were derived from the 1981–2010 climate normal in NCDC's Comparative Climatic Data spreadsheet. Including only those cities with populations greater than 50,000, the top ten are: Mobile; New Orleans; West Palm Beach; Miami; Pensacola; Baton Rouge; Port Arthur, Texas; Tallahassee; Apalachicola; and Wilmington, North Carolina.

23 **The ascending clouds:** Author interview with Mass; and Mass, *The Weather of the Pacific Northwest*, 23–25.

23 **An acre of corn:** U.S. Geological Survey, "The Water Cycle," Transpiration, http://water.usgs.gov/edu/watercycletranspiration.html.

23 **But Yuma is the rain-scarcest city:** NOAA's National Climatic Data Center, Asheville, North Carolina.

23 **Known as rain streamers:** Ahrens, *Meteorology Today*, 473.

24 **Scientists define the monsoon:** Author interview with the geoscientist Peter D. Clift, November 7, 2013.

24 **Almost two-thirds:** Peter D. Clift and R. Alan Plumb, *The Asian Monsoon: Causes, History and Effects* (Cambridge: Cambridge University Press, 2008), 197.

24 **The monsoons:** Ibid., vii.

26 **The anthropologist:** Brian Fagan, *The Long Summer: How Climate Changed Civilization* (New York: Basic Books, 2004), xiii–xiv.

27 **Their subsequent research:** Personal communication with Mark Changizi; and Mark Changizi, R. Weber, R. Kotecha, and J. Palazzo, "Are Wet-Induced Wrinkled Fingers Primate Rain Treads?" *Brain, Behavior and Evolution*, vol. 77, no. 4 (August 2011), http://www.karger.com/Article/FullText/328223#AC.

28 **All in all, Changizi believes:** Personal communication with Changizi; and Changizi, Weber, et al., "Are Wet-Induced Wrinkled Fingers Primate Rain Treads?"

28 **Why would they need:** Chip Walter, *Last Ape Standing: The Seven-Million-Year-Old Story of How and Why We Survived* (New York: Walker Publishing Company, 2013), 9–10.

28 **A new sort of primate:** Ibid., 4.

28 **Emerging paleoclimate:** Gail M. Ashley, "Human Evolution and Climate Change," in Vivien Gornitz, ed., *Encyclopedia of Paleoclimatology and Ancient Environments* (Dordrecht, Netherlands: Springer, 2009), 448–50.

28 **The major leaps:** Interview with Dr. Rick Potts, "The Adaptable Human," *NOVA*, October 26, 2009.

29 **the largest predator:** Christopher A. Brochu and Glenn W. Storrs, "A Giant Crocodile from the Plio-Pleistocene of Kenya, the Phylogenetic Relationships of Neogene African Crocodylines, and the Antiquity of Crocodylus in Africa," *Journal of Vertebrate Paleontology*, vol. 32, no. 3 (2012), 587–602.

29 **Many archaeologists:** Fagan, *The Long Summer*, 19.

31 **Cores from the Arabian Sea suggest:** David E. Anderson, Andrew S. Goudie, and Adrian G. Parker, *Global Environments Through the Quaternary: Exploring Environmental Change* (Oxford: Oxford University Press, 2nd edition, 2013), 127.

31 **As Asian Ice Age hunters:** Wolfgang Behringer, *A Cultural History of Climate* (Cambridge, U.K.: Polity Press, 2010), 33.

31 **After millions of years of heave and ho:** Fagan, *The Long Summer*, xii, 22–25.

31 **Paleolithic people did endure:** David G. Anderson, Albert C. Goodyear, James Kennett, and Allen West, "Multiple Lines of Evidence for Possible Human Population Decline/Settlement Reorganization During the Early Younger Dryas," *Quaternary International*, vol. 242, no. 2 (October 15, 2011), 570–83.

32 **Likewise in Asia:** Clift and Plumb, *The Asian Monsoon*, 200.

32 **Radiocarbon dating lets scientists analyze:** A. M. T. Moore, "The Impact of Accelerator Dating at the Early Village of Abu Hureyra on the Euphrates," *Radiocarbon*, vol. 34, no. 3 (1992), 850–58.

32 **Many scientists believe the faltering rains:** Clift and Plumb, *The Asian Monsoon*, 200–203; and Fagan, *The Long Summer*, 128–45.

TWO: DROUGHT, DELUGE, AND DEVILRY

34 **Nearly 5,000 years ago, the Harappan people:** Richard H. Meadow and Jonathan Mark Kenoyer, "Recent Discoveries and Highlights from Excavations at Harappa: 1998–2000," http://www.harappa.com/indus4/e1.html.

34 **All of this life flourished on monsoon rains and rivers:** Clift and Plumb, *The Asian Monsoon*, 210.

34 **people began to abandon:** Jonathan Mark Kenoyer, *Ancient Cities of the Indus Valley Civilization* (Oxford: Oxford University Press, 1998), 18.

34 **The Hindu holy text:** Author interview, Hindu scholar Vasudha Narayanan,

April 18, 2013; and Vasudha Narayanan, unpublished manuscript, *A Hundred Autumns to Live.*

35 **Appearing in the text more than seventy times:** Clift and Plumb, *The Asian Monsoon,* 210–11.

35 **Searching in an ancient rain-fed lake:** Yama Dixit, David A. Hodell, and Cameron A. Petrie, "Abrupt Weakening of the Summer Monsoon in Northwest India 4,100 Yr Ago," *Geology,* vol. 42, no. 4 (February 24, 2014), 339–42.

35 **The public was captivated:** "The Royal Graves of Ur," British Museum, highlights, http://www.britishmuseum.org/explore/highlights/articles/r/the_royal_graves_of_ur.aspx.

36 **They conceived the first written philosophy:** University of Chicago Library, "This History, Our History: Ancient Mesopotamia," http://mesopotamia.lib.uchicago.edu/mesopotamialife/index.php.

37 **And then, after a hundred years of prosperity:** H. Weiss, M. A. Courty, W. Wetterstrom, F. Guichard, L. Senior, R. Meadow, and A. Curnow, "The Genesis and Collapse of Third Millennium North Mesopotamian Civilization," *Science,* vol. 261 (August 20, 1993), 995–1003.

37 **The city once so grand:** Richard Conniff, "When Civilizations Collapse," *Environment Yale,* http://environment.yale.edu/envy/stories/when-civilizations-collapse/.

37 **With no rain to moisten the soil:** Weiss et al, "The Genesis and Collapse of Third Millennium North Mesopotamian Civilization," 995–1003.

37 **Lake-bed soils in Africa:** H. M. Cullen, P. B. deMenocal, S. Hemming, G. Hemming, F. H. Brown, T. Guilderson, and F. Sirocko, "Climate Change and the Collapse of the Akkadian Empire: Evidence from the Deep Sea," *Geology,* vol. 28 (April 2000), 375–78.

38 **In China, scientists note:** M. J. C. Walker, M. Berkelhammer, S. Bjorck, L. C. Cwynar, D. A. Fisher, A. J. Long, J. J. Lowe, R. M. Newnham, S. O. Rasmussen, and H. Weiss, "Formal Subdivision of the Holocene Series/Epoch," discussion paper, *Journal of Quaternary Science,* vol. 27, no. 7 (2012), 649–59; and Weiss et al., "The Genesis and Collapse of Third Millennium North Mesopotamian Civilization," 995–1003.

38 **More than 11,000 people were killed:** Ahrens, *Meteorology Today,* 453–54.

39 **Persistently copious rains:** Behringer, *A Cultural History of Climate,* 128–29.

39 **Chroniclers of the day:** Ibid., 104.

39 **In 1315, the downpours began:** Henry S. Lucas, "The Great European Famine of 1315, 1316, and 1317," *Speculum,* vol. 5, no. 4 (October 1930), 346.

39 **Floodwaters ran so deep:** Ibid., 346–48.

39 **The Danube burst its banks:** Behringer, *A Cultural History of Climate.*

40 **"The men stood knee-deep:"** Lucas, "The Great European Famine of 1315, 1316, and 1317," 349.

40 **The Flemings thanked God:** Brian Fagan, *The Little Ice Age: How Climate Made History, 1300 to 1850* (New York: Basic Books, 2000), 31–32.

40 **In England, the price of wheat:** Lucas, "The Great European Famine of 1315, 1316, and 1317," 351–52.

40 **In some rural areas:** Fagan, *The Little Ice Age*, 32.

40 **Families foraged:** Lucas, "The Great European Famine of 1315, 1316, and 1317," 355–56.

40 **"Horse meat was precious":** Johannes de Trokelowe, *Annates*, H. T. Riley, ed., Rolls Series, no. 28, vol. 3 (London, 1866), 92–95. Translated by Brian Tierney, Internet Medieval Source Book, Fordham University, http://www .fordham.edu/halsall/source/famin1315a.asp.

40 **They solicited charity:** Fagan, *The Little Ice Age*, 38.

40 **"The people were in such great need":** Ibid. 41.

41 **In Tournai:** John Aberth, *From the Brink of the Apocalypse: Confronting Famine, War, Plague and Death in the Later Middle Ages* (New York and London: Routledge, 2nd edition, 2013), 22.

41 **In Holland, "rich and poor":** Lucas, "The Great European Famine of 1315, 1316, and 1317," 370.

41 **In the rains, dead bodies:** Ibid., 358.

41 **Scholars estimate the Great Famine of 1315–1322 killed some 3 million people:** Ibid., 361–63.

41 **A musician in the papal court of Avignon:** Aberth, *From the Brink of the Apocalypse*, 4.

42 **But the "full world":** Fagan, *The Little Ice Age*, 81.

42 **The stress of childhood hunger:** Behringer, *A Cultural History of Climate*, 107–8.

42 **What we know as:** Fagan, *The Little Ice Age*, 81.

42 **When a victim coughs:** World Health Organization, "Plague," fact sheet, http://www.who.int/topics/plague/en/.

43 **Others who have examined it:** Mark Wheelis, "Biological Warfare at the 1346 Siege of Caffa," *Emerging Infectious Diseases Journal*, vol. 8, no. 9 (September 2002), http://wwwnc.cdc.gov/eid/article/8/9/01-0536 .htm.

43 **By 1349, it had crept north to Scotland:** Fagan, *The Little Ice Age*, 82.

43 **No event in European history:** Behringer, *A Cultural History of Climate*, 109.

43 **A global consortium of scientists:** Nils Christian Stenseth et al., "Plague Dynamics Are Driven by Climate Variation," *Proceedings of the National Academy of Sciences of the United States of America*, vol. 103, no. 35 (August 2006), 13110–15.

43 **At the same time, a good, hard rain:** Molly Caldwell Crosby, *The American Plague: The Untold Story of Yellow Fever, the Epidemic That Shaped Our History* (New York: Berkley Books, 2007).

44 **In late August 1589:** "Annual Report of the Deputy Keeper of the Public Records, Volumes 46–47," Great Britain Public Records, no. 5, Report on Royal Archives of Denmark. This record spells Peter Munch's name as Munck.

44 **King James VI of Scotland had seen the fair Anna:** Agnes Strickland, *Lives of the Queens of England*, vol. VII (London: Henry Colburn, 1844), 323–25.

44 **Munch thought the tempest uncommonly fierce:** Ibid., 326.

44 **The dastardly winds:** Ibid.

45 **After an awkward first embrace:** *Letters to King James the Sixth* (Edinburgh: Library of the Faculty of Advocates, 1835), xvii.

45 **The royal couple's return:** Donald Tyson, *The Demonology of King James I* (Woodbury, Minn.: Llewellyn Publications, 2011), 22.

46 **Historians who study witch hunts:** Wolfgang Behringer, *Witches and Witch-Hunts: A Global History* (Cambridge, U.K.: Polity Press, 2004), 145.

46 **About 80 percent of the victims were female:** Ibid., 37.

46 **A German woodcut from 1486:** Behringer, *A Cultural History of Climate*, figure 4.2, 129.

46 **A frontispiece from a 1489 pamphlet:** Behringer, *Witches and Witch-Hunts*, plate 5, 75.

46 **A colored Swiss painting:** Ibid., plate 6, 84.

47 **The crime disappeared:** Behringer, *A Cultural History of Climate*, 128.

47 **"Witches were the scapegoats":** Ibid.

47 **In 1582, a similar sheet reported:** Behringer, *Witches and Witch-Hunts*, 91.

48 **In Scotland, the zeal for witch-hunting:** Ibid., 6–7.

48 **After humiliations and tortures:** Tyson, *The Demonology of King James I*, 24.

48 **She explained that Satan:** Ibid., 4, 24.

48 **No words could have rung:** Ibid.

49 **Its woodcuts show a storm:** *Newes from Scotland*, Special Collections Exhibition, Special Collections Department, Library, University of Glasgow, Scotland, http://special.lib.gla.ac.uk/exhibns/month/aug2000.html.

49 **She said the witches:** Tyson, *The Demonology of King James I*, 25.

49 **King James and his council:** *Newes from Scotland*.

50 **Witch hunts and trials continued:** Ibid.

50 **He began writing with James in mind:** Stephen Greenblatt, *Will in the World: How Shakespeare Became Shakespeare* (New York: Norton, paperback edition, 2005), 349.

51 **But he relied on storms:** Roland Mushat Frye, *Shakespeare: The Art of the Dramatist* (Boston: Houghton Mifflin Co., 1970), 216.

51 **Archaeologists have found that the later Harappan:** Camilo Ponton, Liviu Giosan, Tim I. Eglinton, Dorian Q. Fuller, Joel E. Johnson, Pushpendra Kumar, and Tim S. Collett, "Holocene Aridification of India," *Geophysical Research Letters*, vol. 39 (February 2012).

THREE: PRAYING FOR RAIN

52 **As a teenager, he'd been stricken:** "Robert McAlpin Williamson," Justices of Texas 1836–1986, University of Texas School of Law Tarlton Law Library Jamail Center for Legal Research, http://tarlton.law.utexas.edu/justices/profile/view/116.

52 **His disability stopped him from nothing:** "Robert McAlpin Williamson," Justices of Texas 1836–1986; and "Williamson, Robert McAlpin (Three Legged Willie)," *Handbook of Texas Online*, Texas State Historical Association, http://www.tshaonline.org/handbook/online/articles/fwi42.

52 **As a brilliant orator:** Ross Phares, *Bible in Pocket, Gun in Hand: The Story of Frontier Religion* (New York: Doubleday & Co., 1962), 133–34.

52 **Tree-ring researchers:** Malcolm K. Cleaveland, Daniel K. Stahle, Todd H. Votteler, Richard C. Casteel, and Jay L. Banner, "Extended Chronology of Drought in South Central, Southeastern and West Texas," Texas Water Forum, http://www.jsg.utexas.edu/ciess/files/Water_Forum_01_Votteler.pdf.

53 **"O Lord, Thou Divine Father":** Phares, *Bible in Pocket, Gun in Hand*, 134.

53 **At Texas A&M University:** John Burnett, "Drought, Wildfires Haven't Changed Perry's Climate-Change Views," NPR News, September 7, 2011; and Kate Sheppard, "Rick Perry Asks Texans to Pray for Rain," *Mother Jones*, April 21, 2011.

54 **His rain refrain:** Timothy Egan, "Rick Perry's Unanswered Prayers," *New York Times*, August 11, 2011.

54 **"trying to co-opt the most important three days of the Christian calendar,"** Richard Connelly, "We Obey Rick Perry, Our Rain Prayer," *Houston Press*, April 22, 2011.

54 **In the arid American Southwest:** Ann Marshall, *Rain: Native Expressions from the American Southwest* (Santa Fe: Museum of New Mexico Press, 2000), 9, 18, 20.

54 **Rain has been woven:** Ibid., 32.

54 **The cantor dons:** Ronald H. Isaacs, *The Jewish Sourcebook on the Environment and Ecology* (Northvale, N.J.: Jason Aronson Inc., 1998), 33.

54 **Today, Muslims communally perform the prayer:** Sarah Kate Raphael, *Climate and Political Climate: Environmental Disasters in the Medieval Levant* (Leiden, Netherlands: Koninklijke Brill, 2013), 70.

54 **In an unprecedented gesture:** Judith Sudilovsky, "Holy Land Jews and Muslims Pray Together for Rain," *Ecumenical News International*, November 17, 2010.

55 **The rain god Iškur/Adad:** Piotr Bienkowski and Alan Millard, *Dictionary of the Ancient Near East* (Philadelphia: University of Pennsylvania Press, 2000), 1–2. You can see Iškur/Adad today in the Louvre. The small figure hardly looks fierce behind his glass pane, but proudly rides his bull across a stone tablet, poised to hurl one of his lightning bolts.

55 **Rain and storm gods were around then, too:** Alberto R. W. Green, *The*

Storm-God in the Ancient Near East, University of California San Diego Biblical and Judaic Studies, Volume 8 (Winona Lake, Indiana: Eisenbrauns, 2003), 76–83.

55 **Some were considered divine kings:** Daniel Schwemer, "The Storm-Gods of the Ancient Near East: Summary, Synthesis, Recent Studies, Part I," *Journal of Ancient Near Eastern Religions,* vol. 7, no. 2 (December 2007), 121–68.

55 **Still thought of as paeans to fertility:** Green, *The Storm-God in the Ancient Near East,* 76–83.

56 **Bulls were a common rain-god motif:** Stephen H. Schneider, Terry L. Root, and Michael D. Mastrandrea, eds., *The Encyclopedia of Climate and Weather,* vol. 1 (New York: Oxford University Press, 2011), "Religion and Weather," 5.

56 **The Aztecs had tried to please:** Isabel De La Cruz, Angelica Gonzalez-Oliver, Brian M. Kemp, Juan A. Roman, David Glenn Smith, and Alfonso Torre-Blanco, "Sex Identification of Children Sacrificed to the Ancient Aztec Rain Gods in Tlatelolco," *Current Anthropology,* vol. 49, no. 3 (June 2008), 519–26.

56 **Jewish tradition identifies:** Isaacs, *The Jewish Sourcebook on the Environment and Ecology,* 155.

56 **In Sanskrit, the word for rain:** Sir Monier Monier-Williams, *A Sanskrit-English Dictionary* (Delhi: Motilal Banarsidass, 2005 reprint; originally published in 1899 by Oxford University Press), 1013.

56 **Hindus consider rivers female:** Vasudha Narayanan, "Water, Wood, and Wisdom: Ecological Perspectives from the Hindu Traditions," *Daedalus: Journal of the American Academy of Arts and Sciences,* vol. 130, no. 4 (Fall 2001), 179–206.

57 **The Australians also attributed:** Schneider et al., *The Encyclopedia of Climate and Weather,* vol. 1: "Religion and Weather," 5.

57 **Other storm gods became known:** David Leeming, *The Oxford Companion to World Mythology* (New York: Oxford University Press, 2005), 401.

57 **One cuneiform tablet describes Iškur:** Douglas R. Frayne, *The Royal Inscriptions of Mesopotamia, Early Periods,* vol. 4: *Old Babylonian Period (2003–1595 BC)* (Toronto: University of Toronto Press, 1990), 271–72.

57 **"Above, Adad made scarce his rain":** Albert T. Clay, *A Hebrew Deluge Story in Cuneiform and Other Epic Fragments in the Pierpont Morgan Library,* Yale Oriental Series: Researches, vol. 5, part 3 (New Haven, Conn.: Yale University Press, 1922), 17–18.

58 **As Methodism's founder, John Wesley, explained:** John Wesley, *Wesley's Notes on the Bible,* "John Wesley's Notes on the Whole Bible, New Testament," Christian Classics Ethereal Library, XXVI, 4 (full text at http://www.ccel.org/ccel/wesley/notes.html).

58 **The newly settled Israelites:** Daniel Hillel, *The Natural History of the Bible: An Environmental Exploration of the Hebrew Scriptures* (New York: Columbia University Press, 2006), 157.

58 "And it shall come to pass": Deuteronomy 11:13–15, 21st Century King James version, www.biblegateway.com.

58 "Take heed to yourselves": Deuteronomy 11:16–17, ibid.

59 "The Lord shall open": Deuteronomy 28:12, ibid.

59 As Rabbi Tanchum bar Chiyya put it: Isaacs, *The Jewish Sourcebook on the Environment and Ecology,* 159.

59 "In the wilderness of the desert": Clift and Plumb, *The Asian Monsoon,* 226.

60 But it is often the wettest: Author interview, Clift; and Clift and Plumb, *The Asian Monsoon,* 223.

60 Krishna's skin is storm-blue: Author interview with Narayanan.

61 In 2001, the Indian government: Ishaan Tharoor, "The World of the Kumbh Mela: Inside the Largest Single Gathering of Humanity," *Time,* January 15, 2013.

61 A crush of devotees waits: Kamakhya Temple, "Ambubachi Mela," http://www.kamakhyatemple.org/Ambubachi.aspx.

61 Mythology surrounding the Saraswati: Author interview with Narayanan.

62 Woolley was not trying to make history: William Ryan and Walter Pitman, *Noah's Flood: The New Scientific Discoveries About the Event That Changed History* (New York: Simon & Schuster, 1998), 53–54.

62 Rainbows exist as a sign of this pact: Genesis chapters 7 through 9, 21st Century King James Version, www.biblegateway.com.

63 He loved the work so much: Norman Cohn, *Noah's Flood: The Genesis Story in Western Thought* (New Haven, Conn.: Yale University Press, 1996), 11–14.

63 Smith was "a highly nervous, sensitive man": Sir E. A. Wallis Budge, *The Rise and Progress of Assyriology* (London: Richard Clay & Sons, 1925), 153.

63 He wrote of the moment: Ryan and Pitman, *Noah's Flood,* 28.

64 According to a colleague's written account: Ibid.

64 The telling is near-exact to Genesis: Ryan and Pitman, *Noah's Flood,* 28.

64 The flood tale got around: Cohn, *Noah's Flood,* 8–9.

65 His 1929 book, *Ur of the Chaldees*: Ryan and Pitman, *Noah's Flood,* 55.

66 It describes a deluge: David R. Montgomery, *The Rocks Don't Lie: A Geologist Investigates Noah's Flood* (New York: W. W. Norton, 2012), 154; and Cohn, *Noah's Flood,* 1.

66 For two decades, Ryan and Pitman have built evidence: Personal communication with William Ryan; Ryan and Pitman, *Noah's Flood*; and K. K. Eris, W. B. F. Ryan, et al., "The Timing and Evolution of Post-glacial Transgression Across the Sea of Marmara Shelf South of Istanbul," *Marine Geology,* vol. 243 (2007), 57–76.

66 The geologists hypothesize: Ryan and Pitman, *Noah's Flood,* 234.

FOUR: THE WEATHER WATCHERS

71 "the Wind encreased": Daniel Defoe and Richard Hamblyn, *The Storm: Edited with an Introduction and Notes by Richard Hamblyn* (London: Penguin, 2005), 26.

71 **Had he been killed:** John J. Miller, "Writing Up a Storm," *Wall Street Journal*, August 13, 2011.

71 **Not a bad guess:** Maximillian Novak, *Daniel Defoe: Master of Fictions: His Life and Ideas* (Oxford: Oxford University Press, 2001), 138.

71 **He was fined two hundred marks:** Charlie Connelly, *Bring Me Sunshine* (London: Little, Brown, 2012), 209.

72 **On the morning of the 27th:** Defoe and Hamblyn, *The Storm*, 34.

72 **Hardly anyone had slept:** Ibid.

72 **Defoe's eyewitness account:** Miller, "Writing Up a Storm."

72 **Defoe placed ads:** Connelly, *Bring Me Sunshine*, 211.

72 **The heart of *The Storm*:** Defoe and Hamblyn, *The Storm*, 64.

73 **Emerging atmospheric science shows up alongside:** Jayne Elizabeth Lewis, *Air's Appearance: Literary Atmosphere in British Fiction, 1660–1794* (Chicago: University of Chicago Press, 2012), 94.

73 **"I cannot doubt but the Atheist's hard'ned Soul":** Defoe and Hamblyn, *The Storm*, 7.

73 **The earliest-known recorded rain science:** Ian Strangeways, *Precipitation: Theory, Measurement and Distribution* (Cambridge: Cambridge University Press, 2007), 6.

73 **"Was there ever a shower":** Aristophanes, "The Clouds," 423 B.C., in *Selected Writings on Socrates* (London: Collector's Library, 2004), 378.

73 **Soaking into the earth:** Malcolm Wilson, *Structure and Method in Aristotle's Meteorologica: A More Disorderly Nature* (Cambridge: Cambridge University Press, 2013), 46, 60.

74 **"In front of the storehouse":** Strangeways, *Precipitation*, 139.

74 **Sejong wanted every village:** Park Seong-Rae, *Science and Technology in Korean History: Excursions, Innovations, and Issues* (Fremont, Calif.: Jain Publishing Co. Asian Humanities Press, 2005), 100.

74 **European weather watchers:** Defoe and Hamblyn, *The Storm*, 27.

74 **But it is no surprise:** Strangeways, *Precipitation*, 141.

74 **In Italy, Evangelista Torricelli:** N. C. Datta, *The Story of Chemistry* (Hyderabad, India: Universities Press India, 2005), 86–87.

75 **Torricelli also gave science:** Gabrielle Walker, *An Ocean of Air: Why the Wind Blows and Other Mysteries of the Atmosphere* (New York: Houghton Mifflin Harcourt, 2007), 8–11.

75 **"All the clouds knew":** Douglas Adams, *So Long, and Thanks for All the Fish* (New York: Del Ray Trade Paperback Edition, 2009), 10.

75 **His least-favorite:** Ibid., 8–9.

75 **Some lexicographers suggest:** Elizabeth Knowles, *The Oxford Dictionary of Phrase and Fable* (New York: Oxford University Press, 2nd edition, 2006), 593.

76 **Linguists mapping dialect:** William A. Kretzschmar Jr., *The Linguistics of Speech* (New York: Cambridge University Press, 2009), 89–92.

77 **Clouds "are commonly as good":** Gavin Pretor-Pinney, *The Cloudspotter's Guide: The Science, History, and Culture of Clouds* (New York: Perigee/Penguin, 2006), 186.

77 **Twice a day:** Richard Hamblyn, *The Invention of Clouds: How an Amateur Meteorologist Forged the Language of the Skies* (New York: Farrar, Straus and Giroux, 2001), 82–83.

77 **Howard's proposed classification:** Ibid., 50.

78 **As the British cloud enthusiast:** Pretor-Pinney, *The Cloudspotter's Guide*, 53.

78 **It is brilliantly simple:** Scott Huler, *Defining the Wind: The Beaufort Scale, and How a 19th-Century Admiral Turned Science into Poetry* (New York: Three Rivers Press, 2004), 8.

79 **The latter was often more thrilling:** Lance Morrow, "The Religion of Big Weather," *Time*, vol. 147, no. 4 (January 22, 1996), 72.

79 **So it was with the man:** "George James Symons, F.R.S.," *Quarterly Journal of the Royal Meteorological Society*, Great Britain, vol. 26, no. 114 (April 1900), 155.

79 **Historians of science:** Katharine Anderson, *Predicting the Weather: Victorians and the Science of Meteorology* (Chicago: University of Chicago Press, 2005), 100.

79 **The first issue:** "George James Symons, F.R.S.," 174.

80 **As they tried to save lives:** Anderson, *Predicting the Weather*, 15.

80 **For one, the large ship-salvage companies:** Malcolm Walker, *History of the Meteorological Office* (Cambridge: Cambridge University Press, 2012), xvii.

80 **By 1865, the British Rainfall Organisation:** "George James Symons, F.R.S.," 155.

81 **He ultimately gathered:** Connelly, *Bring Me Sunshine*, 41–42.

81 **With kind, crinkly eyes:** H. Sowerby Wallis, *British Rainfall 1899: On the Distribution of Rain over the British Isles During the Year 1899* (London, 1900), 18.

81 **He maintained a patient:** Anderson, *Predicting the Weather*, 100.

81 **"Vulgar fractions should never be employed":** Hugh Robert Mill, *British Rainfall 1905: On the Distribution of Rain in Space and Time over the British Isles During the Year 1905*, 45th Annual Volume (London, 1906), 272.

81 **Such a move would undermine:** Anderson, *Predicting the Weather*, 103.

82 **Espy sold the notion:** Mark Monmonier, *Air Apparent: How Meteorologists Learned to Map, Predict, and Dramatize the Weather* (Chicago: University of Chicago Press, 1999), 39.

83 **Abbe designed:** Ibid., 48.

83 **Only thirty years old:** W. J. Humphreys, "Cleveland Abbe, 1838–1916," NOAA History, Giants of Science, http://www.history.noaa.gov/giants/abbe.html.

83 **A petition from the Great Lakes:** Monmonier, *Air Apparent*, 48.

83 **Congress approved:** Ibid., 49.

83 **When he died in 1880:** *Frank Leslie's Illustrated Newspaper*, September 11, 1880, 19.

84 **The real scientist behind:** "Edward Lorenz, Father of Chaos Theory and Butterfly Effect, Dies at 90," *MIT News*, April 16, 2008.

86 **The satellites and supercomputers:** Jason Samenow, "The National Hurricane Center's Strikingly Accurate Forecast for Sandy," *Washington Post*, November 1, 2012.

86 **Yet rain continues to defy Big Data:** Chris Anderson, "The End of Theory: The Data Deluge Makes the Scientific Method Obsolete," *Wired*, June 23, 2008.

86 **National Weather Service statistics:** Nate Silver, "The Weatherman Is Not a Moron," *New York Times*, September 7, 2012.

86 **On July 28, 1997:** James Brooke, "Flash Flood at Colorado Trailer Parks Kills 5 and Injures 40," *New York Times*, July 30, 1997.

86 **Doesken has never forgiven himself:** Author interview with Nolan Doesken, June 26, 2014.

88 **Journalists were eager:** Bill Kovach and Tom Rosenstiel, *Blur: How to Know What's True in the Age of Information Overload* (New York: Bloomsbury USA, 2010), 150.

88 *Niles' Weekly* **ran:** "Rain," *Niles' National Register*, vol. 75, no. 4 (January 24, 1849), 159.

88 **The *New York Times* began:** Paul Martin Lester, *Visual Communication: Images with Messages* (Boston: Wadsworth, 2014), 213.

89 **Like most things in newspapers:** Monmonier, *Air Apparent*, 160.

89 **During the New Deal:** Robert Henson, *Weather on Air: A History of Broadcast Meteorology* (Boston: American Meteorological Society, 2010), 6.

89 **One of the classics:** James C. Fidler obituary, *Austin American Statesman*, May 5, 2007.

89 **"By telephone, telegraph, teletype":** Henson, *Weather on Air*, 7.

89 **New York City's first television weathercast:** Ibid.

90 **Many of the first broadcasters:** Nick Ravo, "Clint Youle, 83, Early Weatherman on TV," obituary, *New York Times*, July 31, 1999.

90 **"The result was TV weather's wildest":** Henson, *Weather on Air*, 11.

90 **Nashville poet-forecaster Bill Williams:** Ibid.

90 **Broadcasting the weather at his hometown station:** Ibid., 3.

91 **Before she became a movie star and sex symbol:** Ibid.

91 **She couldn't tell whether:** Nicola D. Gutgold, *Seen and Heard: The Women of Television News* (Plymouth, U.K.: Lexington Books, 2008), 146.

91 **"A trained gorilla":** Henson, *Weather on Air*, 32.

91 **He dreamed of a twenty-four-hour national cable network:** Frank Batten with Jeffrey L. Cruikshank, *The Weather Channel: The Improbable Rise of a Media Phenomenon* (Boston: Harvard Business School Press, 2002), 39.

92 **For years, federal meteorologists:** Ibid., 70.

92 **Critics dismissed the channel:** Alan Fields, *Partly Sunny: The Weather Junkie's Guide to Outsmarting the Weather* (Boulder, Colo.: Windsor Peak Press, 1995), 40.

92 **Landmark invested $32 million:** Batten and Cruikshank, *The Weather Channel*, 127.

93 **"Heat-wave alert":** Henson, *Weather on Air*, 11.

93 **The idea was for the forecasters:** Craig Wilson, "Sunny Skies over the Weather Channel," *USA Today,* June 7, 2007.

93 **Reporting from the field:** Ibid.

93 **"It was awesome":** Jim Cantore, interview by Karen Herman, Archive of American Television, April 24, 2013.

93 **In 2008, NBC:** Michael J. de la Merced, "Weather Channel Is Sold to NBC and Equity Firms," *New York Times,* July 7, 2008.

93 **The man with the twenty-four-hour weather dreams:** Batten and Cruikshank, *The Weather Channel,* 127.

94 **He called global warming:** Charles Homans, "Hot Air: Why Don't TV Weathermen Believe in Climate Change?" *Columbia Journalism Review,* January 7, 2010.

94 **He joined an estimated quarter:** Leslie Kaufman, "Among Weathercasters Doubt on Warming," *New York Times,* March 29, 2010.

94 *Wheel of Fortune's* **host:** Pat Sajak, "Manmade Global Warming: The Solution," on Ricochet: Conservative Conversation and Community, July 25, 2010, http://ricochet.com/archives/manmade-global-warming-the -solution/.

95 **In 2009, Sealls won a science-reporting award:** Michael Malone, "Climate Change Debate: Locally, It's Still Often Too Hot to Handle," *Broadcasting & Cable,* March 3, 2013.

FIVE: THE ARTICLES OF RAIN

96 **In his 1615 memoir:** M. John R. Loadman, *Analysis of Rubber and Rubber-Like Polymers* (Dordrecht, Netherlands: Kluwer Academic Publishers, 1998), 6.

96 **His exhaustive report:** Ibid., 6.

97 **But neither he nor some of the best:** *The Rubber Age* magazine, published by U.S. Rubber Co., July 10, 1917, 310.

97 **He ended up with elastic:** Thomas Hancock, *Personal Narrative of the Origin and Progress of the Caoutchouc or India-Rubber Manufacture in England* (London: Longman, Brown, Green, Longmans & Roberts, 1857), 3.

97 **Hancock had a mind for mechanics:** Ibid., 2.

98 **in 1795 Glasgow had a dozen:** Thomas Martin Devine, *Exploring the Scottish Past: Themes in the History of Scottish Society* (East Lothian, U.K.: Tuckwell Press, 1995), 111.

98 **He began traveling to Edinburgh:** George Macintosh, *Biographical Memoir of Charles Macintosh, Esq., F.R.S.* (Glasgow: Private printing, W. G. Blackie & Co., 1847), 2–17.

98 **Before Macintosh turned twenty:** J. A. V. Butler, "John Maclean, Charles Macintosh, and an Early Chemical Society in Glasgow," *Journal of Chemical Education,* January 1942, 43.

99 **Macintosh had secret sources:** Macintosh, *Biographical Memoir,* 17.

99 **The elder Macintosh:** Ibid., 118.

99 **Thriving on Scotland's rain and mist:** Oliver Gilbert, *Lichens: Naturally Scottish* (Perth: Scottish Natural Heritage, 2004), 7.

99 **Most cudbear manufacturers:** Macintosh, *Biographical Memoir,* 81–82.

100 **In 1819, Glasgow Gas Works:** Ibid., 82.

100 **Highly flammable, naphtha put the fire in "Greek fire":** Adrienne Mayor, *Greek Fire, Poison Arrows, and Scorpion Bombs: Biological and Chemical Warfare in the Ancient World* (New York: Overlook Duckworth, 2003), 241.

100 **Macintosh heated the brew:** Charles Macintosh, "Specification of His Patent for Water-Proof Double Fabrics," *Glasgow Mechanics' Magazine and Annals of Philosophy, Volume 1* (Glasgow: W. R. McPhun, 1824), 405.

101 **In 1822, he obtained patent number 4,804:** Macintosh, *Biographical Memoir,* 82.

101 **Should he fail:** Ian Miller, "Macintosh Mill, Manchester," Oxford Archaeology, http://thehumanjourney.net/index.php?option=com_content&task=view&id=61&Itemid=115.

101 **Franklin wrote back requesting:** Hancock, *Personal Narrative,* 23.

102 **That would turn out to be Thomas Hancock:** John Loadman and Francis James, *The Hancocks of Marlborough: Rubber, Art, and the Industrial Revolution: A Family of Inventive Genius* (Oxford: Oxford University Press, 2009).

102 **In 1825, Macintosh agreed:** Hancock, *Personal Narrative,* 50.

102 **It was not until Hancock's articles began outselling Macintosh's:** Charles Slack, *Noble Obsession: Charles Goodyear, Thomas Hancock and the Race to Unlock the Greatest Industrial Secret of the Nineteenth Century* (New York: Hyperion, 2002), 65.

102 **In 1831, he made Hancock a partner:** *Mechanics Magazine,* no. 656, March 5, 1836, 470.

103 **Men and women had worn cloaks, capes, and ground-sweeping mantles:** Doreen Yarwood, *Illustrated Encyclopedia of World Costume* (London: B. T. Batsford, 1978), 91–94.

103 **often oiled to deflect rain:** "Clark's Umbrella," in *Discovering Lewis & Clark* (Washburn, N.D.: Lewis and Clark Fort Mandan Foundation, 2003), http://www.lewis-clark.org.

103 **Some doctors were convinced:** Macintosh, *Biographical Memoir,* 84.

103 **"Complaints arising":** Hancock, *Personal Narrative,* 54.

103 **Each stitch acted like a tiny straw:** Ibid., 53.

103 **In the early eighteenth century:** Connelly, *Bring Me Sunshine,* 70–71.

104 **Upper-crust women:** "Clark's Umbrella," in *Discovering Lewis & Clark.*

104 **The slave trader turned abolitionist John Newton:** Connelly, *Bring Me Sunshine,* 73.

104 **Seeing one "battered and ruined":** Ibid., 83.

104 **Connelly celebrates Jonas Hanway:** Ibid., 74–75.

105 **Crusoe describes his umbrella:** Daniel Defoe, *The Adventures of Robinson Crusoe* (London: S. O. Beeton, 1862), 134.

106 **He died three days later:** Nick Holdsworth and Robert Mendick, "Prime

Suspect in Georgi Markov 'Umbrella Poison' Murder Tracked Down to Austria," *The Telegraph*, March 23, 2013.

106 **The earliest artifact of an actual umbrella:** Julia Meech, *Rain and Snow: The Umbrella in Japanese Art* (New York: Japan Society, 1993), 36–38.

106 **In Egypt, the umbrella was associated:** Ibid., 37.

107 **The umbrella was especially significant in China:** Ibid., 37–38.

108 **Meet Mary Anderson:** Clarke Stallworth, "Southern Belle Invented Wiper for Windshield," *Birmingham News*, February 20, 1977.

109 **By 1916, most vehicles:** James Scoltock, "Milestones: Mary Anderson," *Automotive Engineer*, December 2011, 7.

109 **Her patent, too, expired:** Arvids Linde, *Preston Tucker & Others: Tales of Brilliant Automotive Innovators & Innovations* (Dorchester, U.K.: Veloce Automotive Publishers, 2011), 149.

110 **Likewise when army guards:** Hancock, *Personal Narrative*, 55.

110 **Charles Macintosh lived to see:** Ibid.

110 **Goodyear lost a legal battle with Hancock:** Slack, *Noble Obsession*, 235.

111 **a twenty-one-year-old McDonald's advertising secretary:** Jamie Fox, "A Meal Disguised as a Sandwich: The Big Mac," 2009, collected for the Literary and Cultural Heritage Map of Pennsylvania, Pennsylvania Center for the Book, http://pabook.libraries.psu.edu/palitmap/BigMac.html.

111 **Apple's Jef Raskin code-named a secret computer project:** Andy Hertzfeld, *Revolution in the Valley: The Insanely Great Story of How the Mac Was Made* (Sebastopol, Calif.: O'Reilly Media, 2004), 272.

111 **The British slang dictionary:** Jonathan Bernstein, *Knickers in a Twist: A Dictionary of British Slang* (Edinburgh: Canongate Books, 2006), 29.

112 **Their family-owned company:** Robert Gore, induction, National Inventors Hall of Fame, U.S. Patent and Trademark Office, http://www.invent.org/hall_of_fame/1_3_0_induction_gore.asp.

112 **After much trial and flooded-tent error:** Kristin Hostetter, "Gore-Tex, the Fabric That Breathes," *Backpacker*, April 1998, 70.

112 **Companies all over the world:** Ibid.

112 **Daniel was the spoiled youngest:** Author interview with Daniel Dunko, Cumbernauld, Scotland, November 19, 2014.

115 **My tour guide was:** Author interview with Willie Ross, Cumbernauld, Scotland, November 20, 2014.

SIX: FOUNDING FORECASTER

119 **As a boy, Tom Jefferson was drawn to a peak:** Thomas Jefferson, "Head and Heart Letter" to Maria Cosway, October 12, 1786, Paris.

120 **The vista drew the boy:** Thomas Jefferson, "A Memoir of the Discovery of Certain Bones of a Quadruped of the Clawed Kind in the Western Parts of Virginia," *Transactions of the American Philosophical Society*, vol. 4 (1799), 255–56.

120 **Year upon year:** Charles A. Miller, *Jefferson and Nature: An Interpretation* (Baltimore: Johns Hopkins University Press, 1988), 41.

120 **"How sublime to look down":** Jefferson, "Head and Heart Letter."

120 **Even Jefferson's admiring biographer:** Dumas Malone, *Jefferson and His Time: Jefferson the Virginian* (Charlottesville: University of Virginia Press, 1948), 144.

120 **But he ignored his Renaissance idol's:** William Howard Adams, *Jefferson's Monticello* (New York: Abbeville Press, 1983), 51.

120 **More recently, ten of the thirteen original British colonies:** Arthur C. Benke, *Rivers of North America: The Natural History* (Burlington, Mass.: Elsevier Academic Press, 2005), 21.

121 **When it came time to sink a well:** Jack McLaughlin, *Jefferson and Monticello: The Biography of a Builder* (New York: Henry Holt and Co., 1988), 156.

121 **When the first British colonists set out across the Atlantic Ocean:** Karen Ordahl Kupperman, "The Puzzle of the American Climate in the Early Colonial Period," *American Historical Review*, vol. 87, no. 5 (December 1982), 1262.

121 **Temperatures plunged:** Thomas Purvis, *Colonial America to 1763* (New York: Facts on File, 1999), 1.

121 **By Thomas Jefferson's day:** Edwin T. Martin, *Thomas Jefferson: Scientist* (New York: Henry Schuman, 1952), chapters 6, 7, and 8, details Buffon's theories and the responses from the New World, including Jefferson's.

122 **This "theory of degeneracy":** Charles Darwin, *The Origin of Species by Means of Natural Selection* (New York and London: D. Appleton and Co., updated 7th edition, 1915), xiv.

122 **Buffon's influential books:** Martin, *Thomas Jefferson: Scientist*, 155.

122 **America is "overrun with serpents, lizards":** Ibid., 160.

122 **They debated the inaccuracies:** Ibid., 192.

122 **Jefferson invested in rain gauges:** Thomas Jefferson, edited by Edwin Morris Betts, *Thomas Jefferson's Garden Book: 1766–1824* (Charlottesville, Va.: Thomas Jefferson Memorial Foundation, 1999), 69.

123 **And often, they were followed by brilliant sun and the rainbows:** Thomas Jefferson to William Dunbar, January 12, 1801, Thomas Jefferson Papers, Library of Congress.

123 **He estimated average rainfall:** *Thomas Jefferson's Garden Book*, 625. Jefferson made this analysis based on records kept by Colonel James Madison, father of the president, whose rainfall data had the advantage of being collected in the same spot year after year, as opposed to Jefferson's dragging of gauges from Monticello to Philadelphia to Paris to the new federal city of Washington.

123 **Modern meteorologists:** Charlottesville, Virginia, Period of Record General Climate Summary—Precipitation, 1893 to 2012, Southeast Regional Climate Center, University of North Carolina at Chapel Hill.

123 **In London:** South East England Precipitation Averages, 1981–2010, Met Office, United Kingdom.

124 **The architectural historian:** McLaughlin, *Jefferson and Monticello*, 156–57.

125 **Cisterns, ranging from the simplest clay pots:** Larry Mays, George P.

Antoniou, and Andreas N. Angelakis, "History of Water Cisterns: Legacies and Lessons," *Water,* vol. 5, no. 4 (2013), 1923.

125 **He spent years struggling:** "Cisterns," Thomas Jefferson Encyclopedia, http://www.monticello.org/site/house-and-gardens/cisterns.

125 **But, contrary to the evoking:** Peter J. Hatch, *A Rich Spot of Earth: Thomas Jefferson's Revolutionary Garden at Monticello* (New Haven, Conn.: Yale University Press, 2012), 89–90.

126 **In the first modern treatise of architecture:** Leon Battista Alberti, *On the Art of Building in Ten Books* (Boston: MIT Press, 1988), 27.

126 **Rain can warp, swell, discolor:** Stewart Brand, *How Buildings Learn: What Happens After They're Built* (New York: Viking, 1994), 114.

127 **As a general rule, the showier the house:** Ibid., 115–16.

127 **Brand found some 80 percent:** Ibid., 58.

127 **Its owner, Pittsburgh businessman Edgar Kaufmann Sr.:** Ibid.

127 **On a "beautiful little ravine":** Bruce Brooks Pfeiffer, *Wright: Building for Democracy* (Cologne, Germany: Taschen, 2004), 45.

128 **"Dammit, Frank—it's leaking on my desk!":** Meryle Secrest, *Frank Lloyd Wright: A Biography* (Chicago: University of Chicago Press, 1992), 372.

131 **Despite all the hardships:** Adams, *Jefferson's Monticello,* 46–47.

131 **He would marvel:** Charles B. Sanford, *The Religious Life of Thomas Jefferson* (Charlottesville: University of Virginia Press, 1988), 72.

132 **Two years later, he sold:** Gaye Wilson, "Jefferson's Long Look West," Thomas Jefferson Foundation's *Monticello,* Summer 2000.

132 **His official instructions:** Thomas Jefferson, "Instructions to Lewis," in Meriwether Lewis and William Clark, *Original Journals of the Lewis and Clark Expedition, 1804–1806,* vol. 7 (New York: Dodd, Mead & Company, 1905), appendix, 249.

132 **"One wants new words":** Walt Whitman, "Specimen Days," *The Complete Prose Works of Walt Whitman,* vol. 1 (New York: G. P. Putnam's Sons, 1902), 268.

134 **He described the Great Plains:** Major S. H. Long, *Account of an Expedition from Pittsburgh to the Rocky Mountains Performed in the Years 1819, 1820,* vol. III (London: Longman, Hurst, Rees, Orme and Brown, 1823), 236.

134 **"The traveller who shall":** Ibid., 24.

134 **Now a long humid cycle:** John C. Hudson, *Across This Land: A Regional Geography of the United States and Canada* (Baltimore: Johns Hopkins University Press, 2002), 284; and Marc Reisner, *Cadillac Desert: The American West and Its Disappearing Water* (New York: Penguin Books, updated edition, 1993), 35–36.

134 **Anyone willing to head west:** "An Act to Secure Homesteads to Actual Settlers on the Public Domain," Chapter 75, 12 Stat. 392 (1862, repealed 1976).

134 **A line down the middle:** J. W. Powell, *Report on the Lands of the Arid Region of the United States* (Washington: Government Printing Office, 1878), 1–3.

135 **Truth was:** George Cameron Coggins, Charles F. Wilkinson, and John D.

Leshy, *Federal Public Land and Resources Law* (New York: Foundation Press, 2002), 79.

135 **Even in what Powell called:** Powell, *Report on the Lands of the Arid Region*, 3.

135 **"too much planning":** Donald Worster, speaking to NPR's Howard Berkes, NPR's *Water in the West* series, "The Vision of John Wesley Powell," August 26, 2003.

SEVEN: RAIN FOLLOWS THE PLOW

136 **"Sun rose beautiful":** Uriah W. Oblinger to Mattie V. Thomas, October 1–2, 1866, Nebraska State Historical Society.

137 **"Is it raining":** Uriah W. Oblinger to Mattie V. Oblinger, September 29, 1872, Nebraska State Historical Society.

138 **When the conquistador:** Jane Braxton Little, "Saving the Ogallala Aquifer," *Scientific American*, Special Edition, vol. 19, no. 1 (March 2009), 32–39.

139 **"It seems this desert":** Uriah W. Oblinger to Mattie V. Oblinger and Ella Oblinger, February 9, 1873, Nebraska State Historical Society.

139 **Aughey reported on several "facts of nature":** John Francis Freeman, *High Plains Horticulture: A History* (Boulder: University Press of Colorado, 2008), 26.

141 **But, as long as they were blessed:** Uriah W. Oblinger to Mattie V. Oblinger and Ella Oblinger, April 13–18, 1873, Nebraska State Historical Society.

141 **Several ranchers died:** Steven Rinella, *American Buffalo: In Search of a Lost Icon* (New York: Random House, 2008), 176.

141 **Never known was the death toll:** Addison E. Shelton, ed., "The April Blizzard, 1873," *Journal of Nebraska History and Record of Pioneer Days*, vol. III, no. 3 (July–September 1920), 1–2.

141 **It was thought to come:** Jordan Almond, *Dictionary of Word Origins: A History of the Words, Expressions and Clichés We Use* (New York: Citadel Press, 1995), 37–38.

142 **At the *Omaha Republican*:** Shelton, ed., "The April Blizzard, 1873," 1.

142 **Sam "seems some":** Uriah W. Oblinger to Mattie V. Oblinger and Ella Oblinger, April 13–18, 1873.

142 **It wasn't the novelty:** Mattie V. Oblinger to Thomas family, May 19, 1873; and Mattie V. Oblinger to George W. Thomas, Grizzie B. Thomas, and Wheeler Thomas Family, June 16, 1873, Nebraska State Historical Society.

143 **"It seems as though we are destined":** Uriah W. Oblinger to Mattie V. Oblinger and Ella Oblinger, February 9, 1873, Nebraska State Historical Society.

143 **The financier Jay Gould:** Clark C. Spence, *The Rainmakers: American "Pluviculture" to World War II* (Lincoln: University of Nebraska Press, 1980), 6.

143 **"I feel that it is not possible":** Uriah W. Oblinger to Thomas Family, September 26, 1880.

144 **Between 1888 and 1892:** Robert V. Hine and John Mack Faragher, *The American West: A New Interpretive History* (New Haven, Conn.: Yale University Press), 340.

144 **"The farmers helpless":** Stephen Crane, *Crane: Prose and Poetry* (New York: Library of America, 1984), 689.

145 **There, the river finishes:** U.S. Army Corps of Engineers, New Orleans District, "The Mississippi River Basin," http://www.mvn.usace.army.mil/Missions/MississippiRiverFloodControl/MississippiRiverTributaries/MississippiDrainageBasin.aspx.

145 **For millennia, the Mississippi:** Mikko Saikku, *This Delta, This Land: An Environmental History of the Yazoo-Mississippi Floodplain* (Athens: University of Georgia Press, 2005), 36–37, 138.

145 **"Without floods":** Christine A. Klein and Sandra B. Zellmer, *Mississippi River Tragedies: A Century of Unnatural Disaster* (New York: New York University Press, 2014), 37.

145 **In the sixteenth century:** Saikku, *This Delta, This Land,* 140.

145 **Native people:** Ibid., 62.

146 **When French settlers floated down:** Klein and Zellmer, *Mississippi River Tragedies,* 34.

146 **Twelve feet of water:** Ibid., 38–39.

147 **The environmental historian Worster:** Donald Worster, *Rivers of Empire: Water, Aridity, and the Growth of the American West* (New York: Oxford University Press, paperback edition, 1992), 20.

147 **Ten years later:** Susan Scott Parrish, "Faulkner and the Outer Weather of 1927," *American Literary History,* vol. 24, no. 1 (Spring 2012), 35.

147 **He uttered the words:** Klein and Zellmer, *Mississippi River Tragedies,* 67.

148 **"When it rains, it pours":** "Morton Salt: When It Rains, It Pours," *Huffington Post,* March 14, 2012.

148 **Yet another was close:** John M. Barry, "After the Deluge," *Smithsonian,* November 2005.

148 **Gauge readings:** Risk Management Solutions, "The 1927 Great Mississippi Flood 80-Year Retrospective," 2007, 2.

148 **each greater than any seen in the preceding ten years:** Barry, "After the Deluge."

148 **New Orleans broke records:** Klein and Zellmer, *Mississippi River Tragedies,* 67.

148 **floodwaters rose phenomenally:** Saikku, *This Delta, This Land,* 156.

149 **"These heights changed the equations":** John M. Barry, *Rising Tide: The Great Mississippi Flood of 1927 and How It Changed America* (New York: Touchstone, 1997), 40.

149 **Thousands of men worked desperately:** Ibid., 202–3.

149 **"Refugees coming into Jackson":** Ibid., 202.

149 **The dynamiting left scars:** Risk Management Solutions, "The 1927 Great Mississippi Flood 80-Year Retrospective," 6.

150 **Some 637,000 people:** Saikku, *This Delta, This Land,* 159.

150 **The river and its tributaries:** Barry, "After the Deluge."

150 **Barry wrote that the government:** Ibid.

150 **mud-caked and barren:** Risk Management Solutions, "The 1927 Great Mississippi Flood 80-Year Retrospective," 7.

150 **"The removal of forests":** Powell, *Report on the Lands of the Arid Region,* 76.

150 **"It's a man-made disaster":** Barry, "After the Deluge."

151 **Jeffersonian dream of land-based democracy:** Donald Worster, *Dust Bowl: The Southern Plains in the 1930s* (Oxford: Oxford University Press, 2004), 11, 47.

151 **In 1925, Ford Motor Company's:** "Ford Makes 500,000 Tractors," *Engineering News-Record,* vol. 94 (1925), 1078.

151 **Along the Mississippi River:** Worster, *Dust Bowl,* 11.

151 **The Dakotas became as arid as the Sonoran Desert:** Ibid., 12.

152 **Two summers later:** Ibid.

152 **Grasshoppers swarmed:** Ibid.

152 **The winds that nearly blew Uriah:** Ibid., 15.

152 **With neither the prairie grass:** Timothy Egan, *The Worst Hard Time: The Untold Story of Those Who Survived the Great American Dust Bowl* (New York: Houghton Mifflin Co., 2006), 10.

152 **"We watched that thing":** Interview with J. R. Davison, "Surviving the Dust Bowl," *American Experience,* PBS, http://www.pbs.org/wgbh/american experience/features/interview/dustbowl-witness-jr-davison/.

153 **Ships three hundred miles:** Worster, *Dust Bowl,* 13–14.

153 **Midafternoon, the warm air:** Ibid., 18.

153 **"It got so black":** Alan Lomax interview with Woody Guthrie, March 21, 1940, record at American Folklife Center, Library of Congress, transcribed at http://soundportraits.org/on-air/woody_guthrie/transcript.php.

154 **Those hardest hit:** Saikku, *This Delta, This Land,* 157.

154 **"People, since its raining":** "Blind" Lemon Jefferson, "Rising High Water Blues," 1927.

155 **"We had no rain":** Uriah W. Oblinger to Charlie Thomas, October 27, 1896, Nebraska State Historical Society.

EIGHT: THE RAINMAKERS

156 **He had sent ahead a freight car:** Robert St. George Dyrenforth, "Report of the Agent of the Department of Agriculture for Making Experiments in the Production of Rainfall," U.S. Senate, 52nd Congress, First Session, 1891–1892, 5–13.

156 **Dyrenforth, his odd freight:** James Rodger Fleming, *Fixing the Sky: The Checkered History of Weather and Climate Control* (New York: Columbia University Press, 2010), 66.

156 **Sporting the pith helmets:** Photograph III, "The Party," in Dyrenforth, "Report of the Agent," 18.

157 **Also on this first line:** Dyrenforth, "Report of the Agent," 13.

157 **Without enough men:** Ibid., 14.

157 **"The delight, nay, enthusiasm":** John Seelye, "'Rational Exultation': The Erie Canal Celebration," *Proceedings of the American Antiquarian Society*, vol. 94, no. 2 (October 1984), 259.

158 **When the men at the second cannon:** Ronald E. Shaw, *Erie Water West: A History of the Erie Canal, 1792–1854* (Lexington: University of Kentucky Press, 1966), 185.

158 **The smallest villages were determined:** Peter L. Bernstein, *Wedding of the Waters: The Erie Canal and the Making of a Great Nation* (New York: W. W. Norton, 2005), 313.

158 **Clutching the end:** Connelly, *Bring Me Sunshine*, 56–57.

158 **From the time of Plutarch:** Spence, *The Rainmakers*, 22.

159 **By the eighteenth century:** Ibid., 24.

159 **In 1842, the U.S. government had hired its first:** Ibid., 10.

159 **A long curtain of showers:** William B. Meyer, *Americans and Their Weather* (New York: Oxford University Press, 2000), 86.

159 **The idea especially alarmed:** Ibid., 87.

160 **"He might enshroud us in continual clouds":** Senator John J. Crittenden of Kentucky, quoted in Fleming, *Fixing the Sky*, 55.

160 **"I would not trust such a power to this Congress":** Senator Andrew P. Butler of South Carolina, quoted in Meyer, *Americans and Their Weather*, 88.

160 **The only redeeming influence:** Spence, *The Rainmakers*, 7.

160 **John Wesley Powell said as much:** Powell, *Report on the Lands of the Arid Region*, 74–75.

160 **The first champion of the theory:** Spence, *The Rainmakers*, 24.

160 **J. C. Lewis "took note":** J. C. Lewis, "Rain Following the Discharge of Ordnance," *American Journal of Science and Arts*, vol. 32 (November 1861), 296.

161 **"The discharge of heavy artillery":** Ibid.

161 **Fighting in the godforsaken mud:** Spence, *The Rainmakers*, 24.

161 **"This fact was well noticed":** Edward Powers, *War and the Weather, or, the Artificial Production of Rain* (Chicago, 1871; Wisconsin: Delavan, 1890), 152.

162 **two hundred siege guns:** Powers, *War and the Weather*, 200.

162 **The war had been fought:** Spence, *The Rainmakers*, 29.

162 **In 1890, Charles Benjamin Farwell:** Fleming, *Fixing the Sky*, 64.

162 **The chief of the division:** Donald Grebner, Pete Bettinger, and Jacek Siry, *Introduction to Forestry and Natural Resources* (London: Academic Press, 2013), 410.

162 **He complained that he had neither men nor means:** Spence, *The Rainmakers*, 29.

162 **The agent was Robert St. George Dyrenforth:** Ibid., 30.

162 **"A patent lawyer":** Ibid., 30–31.

163 **"They Made Rain":** Ibid., 33–34.

163 **When the real experiments began:** Dyrenforth, "Report of the Agent," 19.

163 **Still, Dyrenforth reported dark clouds:** Ibid.

164 **The farm journalists sometimes witnessed:** Fleming, *Fixing the Sky*, 67, 69.

164 **Instead, they repeated the hype:** Spence, *The Rainmakers*, 34.

164 **"more than one Congressman":** Ibid., 35.

164 **The following year, Congress again ignored:** Ibid., 41.

164 **The nocturnal explosions rained:** Clark C. Spence, "The Dyrenforth Rainmaking Experiments: A Government Venture in Pluviculture," *Arizona and the West*, vol. 3, no. 3 (Autumn 1961), 227.

165 **"The scheme":** Ibid., 228.

165 **When he wrote to U.S. agriculture secretary:** Ibid., 229.

165 **That left people particularly vulnerable:** Spence, *The Rainmakers*, 39–40.

165 **Working in explosives:** Louise Pound, "Nebraska Rain Lore and Rain Making, *California Folklore Quarterly*, vol. 5, no. 2 (April 1946), 135.

166 **On a hot July day:** Ibid.

166 **He claimed he'd been forced:** Spence, *The Rainmakers*, 52.

166 **He took bets:** Ibid., 53.

166 **Melbourne began to charge $500:** Pound, "Nebraska Rain Lore and Rain Making," 137.

166 **He toted it:** Spence, *The Rainmakers*, 55–56.

166 **In Canton:** Ibid., 53.

167 **"Melbourne Causes the Rain to Fall":** Ibid., 60.

167 **Police ruled it a suicide:** Ibid., 62.

167 **After Melbourne left Goodland:** Fleming, *Fixing the Sky*, 86–87.

168 **By 1902, he was dabbling:** Spence, *The Rainmakers*, 80–81.

168 **He got into professional rainmaking:** Fleming, *Fixing the Sky*, 90.

168 **"I simply attract clouds":** Nick D'Alto, "The Rainmakers," *Weatherwise*, vol. 53, no. 5 (September/October 2000), 26–33.

169 **"When it comes to my knowledge":** Fleming, *Fixing the Sky*, 90.

169 **He was soon credited:** D'Alto, "The Rainmakers," 26–33.

169 **Hatfield worked like a madman:** Ibid.

169 **In Hatfield's home:** Ibid.

170 **If Morena topped its banks:** Spence, *The Rainmakers*, 90–91.

170 **On January 27 the Lower Otay Dam burst:** Fleming, *Fixing the Sky*, 94.

170 **The disaster became known as "Hatfield's Flood":** Spence, *The Rainmakers*, 91.

170 **The city lawyers:** Ibid., 92.

171 **"He was anxious to explain":** "Italy Engages Rainmaker," *New York Times*, August 22, 1922, 6.

171 **He made a trip to Honduras:** Spence, *The Rainmakers*, 96.

171 **When the spectacle was over:** Ibid., 94–95.

171 **Hatfield's fans urged:** Ibid., 98.

171 **The headline in the *Washington Post*:** "Charles Hatfield, The Rainmaker, Dies in Obscurity," *Washington Post*, April 16, 1958, C2.

171 **When the ag secretary Jeremiah Rusk:** Jeremiah McLain Rusk, "The Future of American Agriculture," in Albert Shaw, ed., *The Review of Reviews*, February 1893, 331.

172 **Wright had been a key:** Peter L. Jakab, "Aerospace in Adolescence: McCook Field and the Beginnings of Modern Flight Research," in Peter Galison and Alex Roland, eds., *Proceedings of the Evolution of Atmospheric Flight in the Twentieth Century*, Dibner Institute for the History of Science and Technology, MIT, April 4–5, 1997.

173 **"The cloud began to fade away":** Manus McFadden, "Is Rainmaking Riddle Solved?" *Popular Science Monthly*, May 1923, 29.

174 **Grandson of the well-known historian:** John W. Servos, *Physical Chemistry from Ostwald to Pauling: The Making of a Science in America* (Princeton, N.J.: Princeton University Press, 1990), 159.

174 **Bancroft had a particular:** Wilder Dwight Bancroft, *Applied Colloid Chemistry* (New York: McGraw-Hill Book Company, 1921), 2.

174 **He pondered the mechanics of raindrops:** Ibid., 283–86.

174 **It is not clear how:** Spence, *The Rainmakers*, 104.

174 **Infused with sand:** Ibid.

174 **He wrote to Warren:** Fleming, *Fixing the Sky*, 113.

174 **The sand shot out:** L. Francis Warren, "Facts and Plans: Rainmaking—Fogs and Radiant Planes," January 2, 1928.

175 **The head of McCook Field:** "Expect to Spend $10,000,000 for New Buildings on Big Air Site," *Dayton Daily News*, October 29, 1922, 11.

175 **Warren and Bancroft might have cashed in:** Fleming, *Fixing the Sky*, 116.

175 **"The idea of the college professor":** Ibid.

175 **"No use arguing":** "Rainmaking Plans Attacked as Futile," *New York Times*, March 22, 1923, 10.

175 **While the electrified sand:** Spence, *The Rainmakers*, 114–15.

176 **With little to show:** Ibid., 114.

176 **They hauled the portable dynamometer:** Jakab, "Aerospace in Adolescence."

176 **This included aircraft icing:** Fleming, *Fixing the Sky*, 139.

177 **Langmuir and Schaefer decided:** Earrington S. Havens, "History of Project Cirrus," General Electric Report No. RL-756, July 1952, 3–5.

177 **It turns out that supercooled:** Fleming, *Fixing the Sky*, 143.

177 **When Schaefer shared his discovery:** Ibid.

177 **Langmuir, fifty miles away:** Ibid., 146–47.

177 **GE lost no time in asking the military:** Ibid., 147.

177 **The *New York Times* crowed:** "Three-Mile Cloud Made into Snow by Dry Ice Dropped from Plane," *New York Times*, November 15, 1946, 24.

178 **Letters, telegrams, and postcards:** Fleming, *Fixing the Sky*, 148.

178 **Vonnegut experimented:** Ibid., 154.

179 **GE employees were:** Havens, "History of Project Cirrus," 13.

179 **"No chemist, physicist, or mathematician":** Fleming, *Fixing the Sky*, 156.

180 **Ten years later, an advisory committee:** Dianne Dumanoski, *The End of*

the Long Summer: Why We Must Remake Our Civilization to Survive on a Volatile Earth (New York: Three Rivers Press, 2009), 136.

180 **The idea was to flood out roads:** Clyde Edward Wood, *Mud: A Military History* (Washington, D.C.: Potomac Books, 2007), 46.

180 **Operating out of Udorn Royal Air Base:** Fleming, *Fixing the Sky*, 180.

180 **In 1972, Seymour Hersh:** Seymour M. Hersh, "Rainmaking Is Used as Weapon by U.S.," *New York Times*, July 3, 1972, 1.

181 **At congressional hearings:** Peter Braestrup, "Witness Silent on Rain War," *Washington Post*, July 27, 1972, A-21.

181 **The years of drought and storm:** Marquis Childs, "Making War with the Weather," *Washington Post*, July 13, 1976, A-19.

181 **Project Popeye's final report:** Wood, *Mud*, 47.

182 **State water scientists in Utah:** Utah Division of Water Resources, "Cloud Seeding," pamphlet, March 2003.

183 **"There is still no convincing scientific proof":** National Research Council, *Critical Issues in Weather Modification Research* (Washington, D.C.: National Academies Press, 2003), 3.

184 **China spends by far the most:** Xinhua Chinese News Agency, "Better Tech to Boost Weather Manipulation," May 23, 2012, http://news.xinhuanet .com/english/sci/2012-05/23/c_131605614.htm.

184 **Roelof Bruintjes, lead scientist for weather modification:** Author interview with Roelof Bruintjes, National Center for Atmospheric Research, April 30, 2014.

185 **Miamians rowed:** Jay Barnes, *Florida's Hurricane History* (Chapel Hill: University of North Carolina Press, 2007), 175–76.

185 **The GE and military scientists:** "Destroying a Hurricane," *New York Times*, September 10, 1947, 26.

185 **He described a "pronounced modification":** Irving Langmuir, *The Collected Works of Irving Langmuir* (Oxford: Pergamon Press, 1960), 172–73.

186 **"Savannah early this morning":** "1947 October 15: Hurricane Hits Savannah," *Savannah Morning News*, August 8, 2010.

186 **The federal government's chief hurricane forecaster:** Jack Williams and Bob Sheets, *Hurricane Watch: Forecasting the Deadliest Storms on Earth* (New York: Vintage Books, 2011), 161.

186 **Soon after the fateful hurricane:** General Electric Company, *General Electric Review*, vol. 55 (1952), 26.

186 **Hurricane scientists were dismayed:** Williams and Sheets, *Hurricane Watch*, 162.

186 **The quote was a flourish:** "Postal Service Mission and 'Motto,'" U.S. Postal Service, postal history, http://about.usps.com/who-we-are/postal-history/ mission-motto.pdf.

187 **In 2009, the billionaire Microsoft founder:** "The Latest on Hurricane Suppression," Intellectual Ventures, IV Insights Blog, October 30, 2012, http://www.intellectualventures.com/insights/archives/the-latest-on -hurricane-suppression.

NINE: WRITERS ON THE STORM

191 **His primary school was a "bleak mausoleum":** Morrissey, *Autobiography* (London: Penguin Classics, 2013), 8.

192 **A young visitor,** Friedrich Engels, *The Condition of the Working-Class in England in 1844* (Hamburg, Germany: Tradition Classics, 1845).

192 **It was Marr who came:** Johnny Marr, personal biographical sketch, http://www.johnny-marr.com/about.

192 **"a vast company that loves misery":** Michael Azerrad, "Book Review: 'Autobiography' by Morrissey," *Wall Street Journal,* December 6, 2013.

192 **Marr said that the band's:** Johnny Marr, "The Smiths Make Their Top of the Pops Debut," *Guardian,* June 13, 2011.

192 **Their 1984 hit:** "500 Greatest Songs of all Time," *Rolling Stone,* April 7, 2011.

192 **"Teenage depression":** Tom Cardy, "He's Not Miserable Now," *Dominion Post* (Wellington, New Zealand), December 14, 2012.

193 **"His composition that evening was full of raindrops":** Tad Szulc, *Chopin in Paris: The Life and Times of the Romantic Composer* (New York: Da Capo Press, 2000), 212.

194 **The Manchester-born music journalist:** Nabeel Zuburi, "'The Last Truly British People You Will Ever Know': Skinheads, Pakis, and Morrissey," in Henry Jenkins III, Tara McPherson, and Jane Shattuc, eds., *Hop on Pop: The Politics and Pleasures of Popular Culture* (Durham, N.C.: Duke University Press, 2002), 541.

195 **The Doors used those sorts of sound effects,** Stephen K. Valdez, *A History of Rock Music* (Dubuque, Ia.: Kendall/Hunt Publishing Co., 2006), 318.

195 **But the song's well-known rain solo:** Jon Pareles, "Harold Rhodes, 89, Inventor of an Electric Piano," obituary, *New York Times,* January 4, 2001.

195 **The area consistently suffers:** Jim Vleming, "Grays Harbor County Profile," Washington State Employment Security Department, July 2012, https://fortress.wa.gov/esd/employmentdata/reports-publications/regional-reports/county-profiles/grays-harbor-county-profile.

196 **Some ascribe the rise of Cobain:** Pete Prown and Harvey P. Newquist, *Legends of Rock Guitar: The Essential Reference of Rock's Greatest Guitarists* (Milwaukee, Wis.: Hal Leonard, 1997), 242–43.

197 **Anthologies seem to have no end:** See the "Rain" entry in Tessa Kale, ed., *The Columbia Granger's Index to Poetry in Anthologies* (New York: Columbia University Press, 13th edition, 2007), 2126–27.

197 **"It's a sorrowful morning":** Emily Dickinson to Susan Gilbert, February 1852, in Thomas H. Johnson, ed., *The Letters of Emily Dickinson,* vol. 1 (Cambridge, Mass.: Belknap Press of Harvard University Press, 1997), 177.

198 **Dickinson considered rain:** "Everything is so still here, and the clouds are cold and gray—I think it will rain soon—Oh I am so lonely!" she writes to her brother in 1851. Emily Dickinson to Austin Dickinson, October 25, 1851, in ibid., 150.

198 **"We are a rather crestfallen company":** Emily Dickinson to Austin Dickinson, June 8, 1851, in ibid., 110–12.

198 **In other letters, Dickinson describes her horror:** Emily Dickinson to Austin Dickinson, October 5, 1851, in ibid., 140.

199 **Dickinson's dark days:** Christopher H. Ramey and Robert W. Weisberg, "The 'Poetical Activity' of Emily Dickinson: A Further Test of the Hypothesis That Affective Disorders Foster Creativity," *Creativity Research Journal*, vol. 16, nos. 2 & 3 (April 1, 2004), 173–85.

199 **In fiction, rain gives a sense:** Charles Dickens, *The Pickwick Papers* (London: Harmsworth Library, 1905), 681.

200 **Louisa was sure:** Bernard N. Schilling, *The Rain of Years: Great Expectations and the World of Dickens* (Rochester, N.Y.: University of Rochester Press, 2001), 21.

200 **In *The Old Curiosity Shop*:** Ibid., 22.

200 **Snowy London winters:** Bill McGuire, *Global Catastrophes: A Very Short Introduction* (New York: Oxford University Press, 2002), 53.

200 **Lord Byron suggested:** Lucian Boia, *The Weather in the Imagination* (London: Reaktion Books, 2005), 107.

201 **The Earth, and he, were grateful:** Walter Raymond, *The Book of Simple Delights* (London: Hodder & Stoughton, 1906), 129.

202 **Hardy, putting hangman:** Thomas Hardy, "The Three Strangers," in *Wessex Tales* (London: Harper & Brothers, 1896), 35–61; and Thomas Foster, *How to Read Literature Like a Professor* (New York: Harper, 2003), 76.

202 **None worked as perfectly:** Harold Bloom, ed., *Ernest Hemingway's A Farewell to Arms* (New York: Infobase Publishing), 26.

202 **"The fiery exultation":** Edward Lewis Wallant, *The Pawnbroker* (Boston: Harcourt/Harvest Paperbacks, 1978; orig. pub. 1961), 239.

202 **The counterpoint to rain as cleanser:** Foster, *How to Read Literature Like a Professor*, 77.

202 **The sight of herself sullied:** Toni Morrison, *Song of Solomon* (New York: Vintage International, 2004), 312–16.

203 **"At the calendar's gloaming":** Timothy Egan, "The Longest Nights," *New York Times*, January 10, 2013.

204 **It is said that one in ten:** Rosie Goldsmith, "Iceland: Where One in 10 People Will Publish a Book," BBC News, October 13, 2013.

204 **In his four-part "Gifts of Rain":** Elmer Kennedy-Andrews, *Writing Home: Poetry and Place in Northern Ireland, 1968–2008* (Cambridge: D. S. Brewer, 2008), 3, 85.

204 **Rain, gray skies, and lightning:** S. E. Gontarski, ed., *A Companion to Samuel Beckett* (West Sussex, U.K.: Wiley-Blackwell, 2010), 193.

204 **"I mean, it wasn't always as stormy":** Woody Allen, *Radio Days* (Orion Pictures, January 1987).

205 **"is for me an aphrodisiac":** Kristi McKim, *Cinema as Weather: Stylistic Screens and Atmospheric Change* (New York: Routledge, 2013), 109.

205 **The final, rain-kiss scene:** McKim, *Cinema as Weather*, 95.

205 **"People are confined":** Woody Allen and Stig Bjorkman, *Woody Allen on Woody Allen* (New York: Grove Press, revised edition, 2005), 186.

207 **The unconvincing line:** Alasdair Glennie, "Andie MacDowell Defends THAT Four Weddings Line," *Daily Mail*, March 29, 2013.

207 **To conjure a tempest:** Alexander Pope, No. 78, June 10, 1713, in *The Works of Alexander Pope, with Notes and Illustrations by Himself and Others*, vol. 5 (London: Longman, Brown, and Co., etc., 1847 edition), 440–41.

208 **Charles Schulz's Snoopy:** Foster, *How to Read Literature Like a Professor*, 74.

208 **"the rain fell in torrents":** Edward Lytton Bulwer, "Paul Clifford," in *The Works of Edward Lytton Bulwer, Esq., in Two Volumes*, vol. 1 (Philadelphia: E. L. Carey and A. Hart, 1836), 523.

208 **"Wild piles of dark and coppery clouds":** Mark Twain (Samuel Clemens), *The American Claimant* (The Gutenberg Project, 2006, http://www.gutenberg.org/files/3179/3179-h/3179-h.htm), originally published 1892.

209 **"*We all need rain*":** Paul A. Woods, *Morrissey in Conversation: The Essential Interviews* (Medford, N.J.: Plexus, 2007), 156. The emphasis is added.

209 **As the millennium turned:** Marilyn Ann Moss, *Raoul Walsh: The True Adventures of Hollywood's Legendary Director* (Lexington: University Press of Kentucky, 2011), 178. Clark Gable and Carole Lombard bought their Encino estate from the director Raoul Walsh. It was on Petit Street, but they changed the name to Tara Drive.

209 **He rented a little guest house:** Budd Schulberg, "Remembering Scott," in Jackson Bryer, Alan Margolies and Ruth Prigozy, eds., *F. Scott Fitzgerald: New Perspectives* (Athens: University of Georgia Press, 2004), 11–12. Scott rented the Encino guest cottage of Edward Everett Horton.

209 **The estate lies just south of Ventura:** 4543 Tara Drive, Encino, California, 91436. Lynn Barber, "The Man with the Thorn in His Side," *Observer*, September 14, 2002. This was where Morrissey was living in 2002, when Lynn Barber of *The Observer* asked him about his estate, and whether he really lived next door to Johnny Depp. Morrissey corrected her: "No—he lives next door to me."

TEN: THE SCENT OF RAIN

210 **While color and song eventually returned:** Waldemar Hansen, *The Peacock Throne: The Drama of Mogul India* (Delhi: Motilal Banarsidass, 1986), 113.

211 **A seven-spouted silver hydra:** Ibid., 124–25.

211 **"Smell," wrote Helen Keller:** Helen Keller, *The World I Live In* (New York: The Century Co., 1904), 66.

212 **"My little friends and I":** Ibid., 67–68.

212 **Geosmin is the bane:** Mark Waer, Leland Harms, Rick Bond, and Nicholas Burns, "A Matter of Taste: How to Control the Odor and Flavor of Water Before Residents Raise a Stink," *American City & County*, June 1, 2003, 16–24.

213 **In the desert Southwest:** Gary Paul Nabhan, *The Desert Smells Like Rain: A Naturalist in O'Odham Country* (Tucson: University of Arizona Press, 1982), 5–6. Nabhan once asked a Papago Indian boy to describe what the desert smelled like to him. "The desert smells like rain," he replied. How could that be, Nabhan wanted to know, when it hardly ever rains there? It turns out that his question had triggered in the child the strong memory of a scent. When the boy pondered the smell of the desert, he remembered feeling overtaken by the odor of creosote bushes after a storm—their aromatic oils released by the rains. The scent and the memory attached to it were so strong that they became, for him, the desert's essence.

213 **"Clean but funky":** Thomas Wolfe, *Look Homeward, Angel* (New York: Scribner, 2000; orig. pub. Charles Scribner's Sons, 1929), 69.

213 **To Sanjiv Chopra:** Deepak Chopra and Sanjiv Chopra, *Brotherhood: Dharma, Destiny, and the American Dream* (Seattle: Amazon Publishing, 2013), 23.

213 **At fifty-three, he wrote of how:** Rafael E. Lopez-Corvo, *The Dictionary of the Work of W. R. Bion* (London: H. Harnac, 2003), 2–3.

213 **The scent can so tantalize:** I. J. Bear and R. G. Thomas, "Nature of Argillaceous Odour," *Nature,* vol. 201, no. 4923 (March 7, 1964), 993–95.

213 **A leading Aussie poet:** Les A. Murray, "February: Feb," from *Les Murray Selected Poems* (Melbourne: Black Inc., 2007), 108.

214 **Using steam distillation:** Bear and Thomas, "Nature of Argillaceous Odour," 993–95.

214 **Just like perfume:** "Petrichor: Rain's Piquant Perfume," *Ecos,* Commonwealth Scientific and Industrial Research Organisation, Australia, February 1976, 32.

214 **Ultimately, Bear and Thomas linked:** I. J. Bear and R. G. Thomas, "Genesis of Petrichor," *Geochimica et Cosmochimica Acta,* vol. 30, no. 9 (September 1966), 869–79.

214 **They help form the blue haze:** William Alyn Johnson, *Invitation to Organic Chemistry* (Sudbury, Mass.: Jones and Bartlett Publishers, 1999), 261–62.

214 **The aroma is:** Bear and Thomas, "Genesis of Petrichor," 869–79.

214 **They called it "petrichor":** Bear and Thomas, "Nature of Argillaceous Odour," 993–95.

225 **Despite geosmin and ozone:** Author interview with Heather Sims, head perfumer, Arylessence, August 22, 2013.

225 **Many of these are behind the most famous perfumes:** Luca Turin, *The Secret of Scent: Adventures in Perfume and the Science of Smell* (New York: Harper Perennial, 2007), 52–53.

226 **Chanel was said to abhor:** Lisa Chaney, *Coco Chanel: An Intimate Life* (New York: Viking, 2011), 185.

226 **Beaux said he chose:** William Lidwell and Gerry Manasca, *Deconstructing Product Design: Exploring the Form, Function, Usability, Sustainability, and Commercial Success of 100 Amazing Products* (Minneapolis: Rockport Publishers, 2011), 130.

ELEVEN: CITY RAINS

228 **South Florida's water supply:** Elizabeth D. Purdum, "Florida Waters," Institute of Science and Public Affairs, Florida State University, April 2002.

229 **It happened that California was searing:** "The 1976–1977 California Drought: A Review," California Department of Water Resources, May 1978.

229 **Historians have variously described:** See eminent historians of both California and Florida who have written histories with the title *Land of Sunshine*: William Deverell and Greg Hise's *Land of Sunshine: An Environmental History of Metropolitan Los Angeles* (Pittsburgh: University of Pittsburgh Press, 2005); and Gary Mormino's *Land of Sunshine, State of Dreams: A Social History of Modern Florida* (Gainesville: University Press of Florida, 2008).

230 **In the nineteenth, flood-weary Mexicans:** Jared Orsi, *Hazardous Metropolis: Flooding and Urban Ecology in Los Angeles* (Berkeley: University of California Press, 2004), 14–15

230 **In L.A. alone:** Kevin Roderick, "Deadly Flood of 1938 Left Its Mark on Southland," *Los Angeles Times,* October 20, 1999.

231 **grizzly bears lumbered down from the hills . . . sewage-treatment plants upstream:** Roderick, "Deadly Flood of 1938"; and Peter J. Westwick and Peter Neushul, *The World in the Curl: An Unconventional History of Surfing* (New York: Crown Publishing Group, 2013), 173–74.

231 **An estimated 85 percent of Los Angeles:** Author interview with Hadley Arnold, March 31, 2014.

231 **Elaborate charts:** California Office of Environmental Health Hazard Assessment, "Safe Eating Guidelines for Fish from Coastal Areas of Southern California: Ventura Harbor to San Mateo Point," http://oehha.ca.gov/fish/so_cal/socal061709.html.

231 **"The water will have":** Ian Lovett, "Slaking a Region's Thirst While Cleaning Its Beaches," *New York Times,* April 7, 2013.

232 **The late writer and Everglades champion:** Jack E. Davis, *An Everglades Providence: Marjory Stoneman Douglas and the American Environmental Century* (Athens: University of Georgia Press, 2009), 522.

232 *devastating, ruining, havoc-wreaking* **rains:** Michael Grunwald, *The Swamp: The Everglades, Florida, and the Politics of Paradise* (New York: Simon & Schuster, 2006), 223.

232 **After the 1947 wet season:** Ibid., 221.

232 **As with the Los Angeles River:** Cynthia Barnett, *Blue Revolution: Unmaking America's Water Crisis* (Boston: Beacon Press, 2011), 29.

234 **When I visited the Miami:** Author interview with Virginia Walsh, Senior Professional Geologist, Chief Hydrogeology Section, Miami Dade Water and Sewer Department, May 28, 2014.

234 **Scientists say Miami:** Southeast Regional Climate Change Compact, "A Unified Sea Level Rise Projection for Southeast Florida," April 2011, iv.

234 **The water may not wash over flood gates as we imagine:** Coral Davenport, "Rising Seas," *New York Times*, March 27, 2014.

235 **This rain runoff:** State of Washington, Puget Sound Partnership, www.psp.wa.gov.

236 **But the greatest tragedy:** Mass, *The Weather of the Pacific Northwest*, 18.

236 **Fleming died:** Staff, "Seattle Audio-book 'Star' Among Four Dead in Storm," *Seattle Times*, December 15, 2006.

236 **Seattle is at the forefront:** Office of the City Clerk, Executive Order: 2013-01 Citywide Green Stormwater Infrastructure Goal & Implementation Strategy, March 6, 2013.

236 **The first green street:** Seattle Public Utilities, Green Stormwater, Street Edge Alternatives, http://www.seattle.gov/util/MyServices/DrainageSewer/Projects/GreenStormwaterInfrastructure/CompletedGSIProjects/StreetEdgeAlternatives/index.htm.

237 **"Be a light":** Denise Whitaker, "Storm Water Pond Dedicated in Memory of Woman Who Drowned," KOMO News, May 22, 2013, http://www.komonews.com/news/local/Storm-water-pond-dedicated-in-memory-of-woman-who-drowned-in-basement-208538021.html.

239 **In spring 2014:** Ed Fletcher, "Folsom Lake's Decline Exposes Gold Rush History," *Sacramento Bee*, December 31, 2013.

239 **Governor Jerry Brown:** Carla Marinucci, "Gov. Jerry Brown Sails into History Books," *SFGate*, September 29, 2013.

240 **During a rare deluge:** "With Rain in the Forecast, LADWP Urges Customers to Turn Off Sprinklers and Save Water," Los Angeles Department of Water and Power, February 25, 2014.

240 **Angelenos use far less:** Paul Rogers and Nicholas St. Fleur, "California Drought: Database Shows Big Difference Between Water Guzzlers and Sippers," *San Jose Mercury News*, February 7, 2014.

240 **Hundreds of acres:** Stephanie Pincetl and Tim Papandreou, "Los Angeles, the Improbable Sustainable City," California Center for Sustainable Communities at UCLA, sustainablecommunities.environment.ucla.edu/2011/09/los-angeles-the-improbable-sustainable-city/.

241 **Bolstered by a cadre of young conservation-design gurus:** See the nonprofit Watershed Management Group, watershedmg.org.

241 **If we do not lower the carbon emissions that are warming the Earth:** U.S. Environmental Protection Agency, "Climate Impacts in the Southwest," http://www.epa.gov/climatechange/impacts-adaptation/southwest.html.

241 **The city has launched an ambitious retrofit:** Author interview with Mark Hanna, September 29, 2014; and City of Los Angeles Stormwater Program, "LADWP Announces Stormwater Capture Master Plan," March 28, 2014.

242 **"The work on the river":** Author interview with Hadley Arnold, Executive Director, Arid Lands Institute at Woodbury University, March 31, 2014.

TWELVE: STRANGE RAIN

247 **"were afraid to move in case we trod on them"**: Anecdote based on first-person accounts recorded by Paul Simons, "Raining Frogs and Monsters: Tornado-like Waterspouts May Explain Showers of Fish and Snails and Other Strange Downpours. But Do They Hold a Clue to the Mystery of Nessie?" *The Guardian*, November 11, 1993; Michael Allaby, *Tornadoes* (New York: Facts on File, 2004), 113; and Bob Rickard and John Michell, *The Rough Guide to Unexplained Phenomena* (New York: Penguin, 2007), 25.

247 **When the rain let up:** David Wallechinsky and Amy Wallace, *The New Book of Lists: The Original Compendium of Curious Information* (New York: Canongate, 2005), 566.

248 **"In Paeonia and Dardania, it has, they say, before now rained frogs"**: Athenaeus, *The Deipnosophists. Or Banquet of the Learned of Athenaeus*, vol. 2 (London: Henry G. Bohn, 1854), 526–27, full text accessed via University of Wisconsin digital library, http://digicoll.library.wisc.edu/cgi-bin/Literature/Literature-idx?id=Literature.AthV2.

248 **In 1946, the professional skeptic:** Bergen Evans, *The Natural History of Nonsense* (New York: Alfred A. Knopf, 1946), 25; and Jerry Dennis, *It's Raining Frogs and Fishes: Four Seasons of Natural Phenomena and Oddities of the Sky* (New York: Harper Perennial, 1993), 47.

248 **Reports of frog rain continue:** Liwei Fu, "Strange Storms: Frogs, Spiders and Fish," *Epoch Times*, story updated January 6, 2011, http://www.theepochtimes.com/n2/science/strange-storms-frogs-fish-insects-from-skies-6468.html.

249 **In 2010, hapless frogs fell:** Kathryn Quinn, "Raining Frogs," *Romanian Times*, June 22, 2010.

249 **The same year, frogs and fish:** Cosmas Butunyi, "Kenya: Experts Warn of More 'Fish and Frogs' Rain," *The Nation*, October 5, 2010.

249 **"I have personally never been so fortunate"**: E. W. Gudger, "Do Fishes Fall from the Sky?" *The Scientific Monthly*, vol. 29, no. 6 (December 1929), 526.

250 **The sky rained mud:** Joe Hasler, "Weird Stories of Objects Falling from the Sky—Explained," *Popular Mechanics*, September 17, 2009.

250 **Australian scientists have some of the best:** Peter J. Unmack, "Biogeography of Australian Freshwater Fishes," *Journal of Biogeography*, vol. 28, no. 9 (September 1, 2001), 1065.

250 **A tornado reported:** Chris Dolce, "Where Frogs, Fish and a Cow Fell from the Sky," Weather.com, August 23, 2013, http://www.weather.com/news/tornado-central/frogs-fish-raining-down-20130411?pageno=2.

250 **The strange rain of Labor Day 1969:** "Charlotte Rains Golf Balls," *St. Petersburg Times*, September 3, 1969.

250 ***Popular Mechanics* magazine speculated:** Hasler, "Weird Stories of Objects Falling from the Sky—Explained."

251 **Those included frog and fish falls:** Charles Fort, *The Book of the Damned: The Collected Works of Charles Fort* (New York: Penguin, 2008; orig. pub. 1919), 29.

252 **Perhaps there was an invisible:** Ibid., 90.

252 **Many of Fort's pet phenomena:** John S. Lewis, *Rain of Iron and Ice: The Very Real Threat of Comet and Asteroid Bombardment* (New York: Basic Books, 1996), 15.

252 **The biologist and bee expert:** Thomas D. Seeley, *Honeybee Democracy* (Princeton, N.J.: Princeton University Press, 2010), 59.

252 **He later worked with the Harvard chemical-weapons expert:** Tomas D. Seeley et al., "Yellow Rain," *Scientific American*, vol. 253, issue 3 (September 1985), 128–37.

252 **Other scientists and former CIA agents:** Paul Hillmer, *A People's History of the Hmong* (St. Paul: Minnesota Historical Society Press, 2010), section 5.

252 **Fort could have told them:** Fort, *The Book of the Damned*, 63.

252 **"My own impositivist acceptances":** Ibid., 36–37.

253 **Louis and Kumar hypothesized:** Godfrey Louis and A. Santhosh Kumar, "The Red Rain Phenomenon of Kerala and Its Possible Extraterrestrial Origin," *Astrophysics and Space Science*, vol. 302, no. 1 (April 2006), 175–87.

253 **When exposed to the extreme heat:** Rajkumar Gangappa, Milton Wainwright, A. Santhosh Kumar, and Godfrey Louis, "Growth and Replication of Red Rain Cells at 121°C and Their Red Fluorescence," Conference Proceedings, vol. 7819, Instruments, Methods, and Missions for Astrobiology XIII, September 2010.

254 **Some children wandering:** Chares Pellegrino, *The Last Train from Hiroshima: The Survivors Look Back* (New York: Henry Holt and Co., 2010), 29.

254 **It would take survivors years:** Lawrence Langer, *Holocaust Testimonies: The Ruins of Memory* (New Haven, Conn.: Yale University Press, 1991), 3.

254 **It was some of the first proof:** Ranjeet S. Sokhi, ed., *World Atlas of Atmospheric Pollution* (London: Anthem Press, 2011), 16.

255 **Given the exact same amount of water vapor:** Ahrens, *Meteorology Today*, 134–35.

255 **In the countryside, he wrote:** Charles Dickens, *Our Mutual Friend,* quoted in Peter Thorsheim, "Interpreting the London Fog Disaster of 1952," in Erna Melanie DuPuis, ed., *Smoke and Mirrors: The Politics and Culture of Air Pollution* (New York: New York University Press, 2004), 155.

255 **When a pea-souper claimed 1,150 lives:** Ahrens, *Meteorology Today,* 462.

256 **People, press, and politicians:** Peter Thorsheim, "Interpreting the London Fog Disaster of 1952" in DuPuis, *Smoke and Mirrors,* 158.

256 **Ultimately, the Great Fog killed 12,000 people:** Christopher Stevens, "The Pea Souper That Killed 12,000: How the Great Smog Choked London 60 Years Ago This Week," *Daily Mail,* December 5, 2012.

256 **Facing intense public pressure:** Thorsheim, "Interpreting the London Fog Disaster of 1952" in DuPuis, *Smoke and Mirrors,* 160.

257 **Not until the Clean Air Act:** American Meteorological Society, "A Look at U.S. Air Pollution Laws and Their Amendments: 1955, 1963, 1970, 1990," http://www.ametsoc.org/sloan/cleanair/cleanairlegisl.html.

257 **Engineers at the Metropolitan Transit Authority:** P. Aarne Vesilind and Thomas D. DiStefano, *Controlling Environmental Pollution: An Introduction to the Technologies, History, and Ethics* (Lancaster, Pa.: DEStech Publications, 2006), 335.

257 **Black rains and black snow:** R. D. Gupta, *Environmental Pollution: Hazards and Control* (New Delhi, India: Concept Publishing Company, 2006), 23–24.

257 **Residents reported that the rains:** Fu Jianfeng, "The Black Rain in Shenzhen," *East-South-West-North,* August 24, 2007, http://www.zonaeuropa .com/20070826_1.htm.

257 **For nearly a century:** Chris C. Park, *Acid Rain: Rhetoric and Reality* (Oxford: Methuen and Co., 1987; citation is to the U.S. edition, New York: Routledge, 2013), 6.

257 **In the 1960s, they began to reveal themselves in Germany's Black Forest:** Ibid., 100–101.

258 **Some streams:** Christine Alewell et al., "Are There Signs of Acidification Reversal in Freshwaters of the Low Mountain Ranges in Germany?" *Hydrology and Earth System Sciences,* vol. 5, no. 3 (2001), 368.

258 **These drops, in turn:** Ahrens, *Meteorology Today,* 483.

258 **The sickest areas:** Park, *Acid Rain,* 3.

259 **In the 1960s, American scientists sampling rainfall:** Gene E. Likens and F. Herbert Bormann, "Acid Rain: A Serious Regional Environmental Problem," *Science,* June 14, 1974, 1176–79.

259 **In an article in the journal *Science*:** Ibid.

259 **They sounded the alarm:** Ibid.

259 **Today it's well under 8.9 million:** U.S. Environmental Protection Agency, "Acid Rain and Related Programs: 15 Years of Results 1995 to 2009," October 2010.

259 **In places most prone:** Meg Marquardt, "Neutralizing the Rain: After Much Success in the Battle Against Acid Rain, Challenges Remain," *Earth,* June 2012.

260 **Foresters believe:** Charles Driscoll et al., "Acid Rain Revisited: Advances in Scientific Understanding Since the Passage of the 1970 and 1990 Clean Air Act Amendments," Hubbard Brook Research Foundation, 2001.

260 **Lakes are healing:** Kristin Waller et al., "Long-term Recovery of Lakes in the Adirondack Region of New York to Decreases in Acidic Deposition," *Atmospheric Environment,* vol. 46 (January 2012), 56–64.

260 **Back at the Hubbard Brook:** Hubbard Brook Experimental Forest, Acid Precipitation Trends, www.hubbardbrook.org.

260 **Scientists who track contaminants in the atmosphere:** National Atmospheric Deposition Program, http://.nadp.sws.uiuc.edu/.

260 **The largest premodern statue:** Christina Larson, "China Takes First

Steps in the Fight Against Acid Rain," *Yale Environment 360,* October 28, 2010.

260 **At the turn of the century, China surpassed the United States:** Zifeng Lu et al., "Sulfur Dioxide and Primary Carbonaceous Aerosol Emissions in China and India, 1996–2010," *Atmospheric Chemistry and Physics,* vol. 11, September 2011, 9848.

260 **In South Florida, studies have shown the massive drainage:** Curtis H. Marshall, Roger A. Pielke Sr., Louis T. Steyaert, and Debra A. Willard, "The Impact of Anthropogenic Land-Cover Change on the Florida Peninsula Sea Breezes and Warm Season Weather," *Monthly Weather Review,* vol. 132 (January 2004), 51.

261 **At the other extreme:** Matt McGrath, "Reservoirs Make Local Flooding Worse, Says Study," BBC News, December 14, 2012.

261 **"As we change":** Dr. Jerad Bales, U.S. Geological Survey, "Water, Energy, and Food Production: Observing, Understanding, Forecasting," University of Florida Water Institute Symposium, remarks and Q&A, February 11, 2014.

261 **Scientists analyzing precipitation:** Robert K. Kaufmann, Karen C. Seto, Annemarie Schneider, Zouting Liu, Liming Zhou, and Weile Wang, "Climate Response to Rapid Urban Growth: Evidence of a Human-Induced Precipitation Deficit," *Journal of Climate,* vol. 20, May 15, 2007, 2299–306.

261 **In the United States:** J. Marshall Shepherd, Harold Pierce, and Andrew J. Negri, "Rainfall Modification by Major Urban Areas: Observations from Spaceborne Rain Radar on the TRMM Satellite," *Journal of Applied Meteorology,* vol. 41 (July 2002), 689–701.

261 **Skyscrapers can have:** Author interview with Bob Bornstein, San Jose State University Department of Meteorology and Climate Science, July 29, 2014.

262 **how the emissions we send:** Global Carbon Project, Carbon Budget 2014, http://www.globalcarbonproject.org/carbonbudget/.

263 **"believed in 'Megonia'":** Steven Gaydos, "Fort, 'Wild Talents' Were Major Influences on Anderson," *Variety,* February 7, 2000.

263 **Because they require both land and aquatic habitat:** Kerry Kriger, "Save the Frogs!" http://www.savethefrogs.com/why-frogs/index.html.

263 **After the nuptials:** Naresh Mitra, "Croaking Consorts in Rain Call," *Times of India,* April 10, 2013.

263 **Among the Zuni:** Ann Marshall, *Rain: Native Expressions from the American Southwest* (Santa Fe: Museum of New Mexico Press, 2000), 69.

263 **Nineteenth-century science journals:** David P. Badger, *Frogs* (Stillwater, Minn.: Voyageur Press, 1995), 23.

263 **Frogs have a "barometric propensity":** Book review of St. George Mivart's *The Common Frog* (London: Macmillan and Co., 1875), *Quarterly Journal of Science and Annals of Metallurgy, Engineering, Industrial Arts, Manufactures, and Technology,* vol. 5 (January 1875), 99.

263 **In Louisiana, the Creole people:** Lafcadio Hearn, *"Gombo Zhèbes": Little*

Dictionary of Creole Proverbs, Selected from Six Creole Dialects (New York: Will H. Coleman, 1885), 22.

264 **The weatherwise chorus:** Badger, *Frogs,* 72.

264 **Those that lay their eggs:** Archie Carr, *A Naturalist in Florida: A Celebration of Eden* (New Haven, Conn.: Yale University Press, 1996), 183.

264 **Nearly two hundred frog species have vanished:** David B. Wake and Vance T. Vrendenburg, "Are We in the Midst of the Sixth Mass Extinction? A View from the World of Amphibians," *Proceedings of the National Academy of Sciences of the United States of America,* colloquium paper, vol. 105, issue supp. 1, 11466.

THIRTEEN: AND THE FORECAST CALLS FOR CHANGE

265 **Tzilkowski and Hixson:** Colleen O'Connor, "Massive Dust Storms Hit Southeast Colorado, Evoking 'Dirty Thirties,'" *Denver Post,* June 9, 2013.

265 **Similar dust storms:** "Despite Fall Floods, Drought Persists in Southeastern Colorado," NOAA, February 18, 2014.

265 **The tumbleweeds have to be pushed:** Garrison Wells, "Colorado Tumbleweeds Explosion Creating Hazards and Headaches for Many," *The Gazette,* February 23, 2014.

266 **In the largest airlift since Hurricane Katrina:** Jason Samenow, "Colorado's 'Biblical' Flood by the Numbers," *Washington Post,* September 16, 2013.

266 **The Boulder area broke every rainfall record on its books:** Charlie Brennan, "Forum: Colorado Flood Not 'The Big One,' but Still an Event for the History Books," *Daily Camera,* February 27, 2014.

266 **A thunderstorm lifting:** "Colorado Remembers Big Thomson Canyon Flash Flood of 1976," NOAA, July 30, 2001.

266 **A stationary low-pressure system:** "Severe Flooding on the Colorado Front Range: A Preliminary Assessment," CIRES Western Water Assessment, University of Colorado, September 2013.

267 **"Rainfall and other weather events":** Martin P. Hoerling, "Climate Change and Extreme Weather: Recent Events and Future Forecasts," Ohio State University Climate Change Webinar, April 24, 2014.

267 **The scientists predict greater intensity and frequency:** Working Group 1, contribution to the Fifth Assessment Report of the Intergovernmental Panel on Climate Change, "Climate Change 2013: The Physical Science Basis, Summary for Policymakers," 5.

268 **As the world consumes:** NASA Earth Observatory, "How Much More Will Earth Warm?" Global Warming, 5, http://earthobservatory.nasa.gov/Features/GlobalWarming/page5.php.

268 **Some of them, including James Hansen:** James Hansen, *Storms of My Grandchildren: The Truth About the Coming Climate Catastrophe and Our Last Chance to Save Humanity* (New York: Bloomsbury USA, 2009), 225–26.

269 **Along the beaches:** Claire Marshall, "Storms and Floods Unearth Unexploded Wartime Bombs," BBC News, February 28, 2014.

269 **The problem for climate scientists:** Thomas C. Peterson, Martin P. Hoerling, Peter A. Stott, and Stephanie C. Herring, "Explaining Extreme Events of 2012 from a Climate Perspective," *Bulletin of the American Meteorological Society,* vol. 94, no. 9 (September 2013), 1.

270 *What constitutes a rainy day?*: G. J. Symons, *British Rainfall 1866: On the Distribution of Rain over the British Isles During the Year 1866, as Observed at Above 1000 Stations in Great Britain and Ireland* (London: Edward Stanford, 1867), 4.

270 **Or he'd go on a tear:** Ibid., 5.

271 **"As well as looking forward":** Author interview with Mark McCarthy, the Met, Exeter, U.K., November 21, 2013.

271 **At *Slate*:** Charlie Warzel, "The Search for the Internet's Next Top Weather Nerd," BuzzFeed, February 5, 2014.

272 **"My wife and I realized":** Eric Holthaus, "Why I'm Never Flying Again," *Quartz,* October 1, 2013.

272 **"What can all this be?":** Defoe and Hamblyn, *The Storm,* 7.

272 **Every major:** National Aeronautics and Space Administration, "Consensus: 97% of Climate Scientists Agree," Global Climate Change website, http://climate.nasa.gov/scientific-consensus.

273 **will come to be seen like the vulture ship-salvagers:** Malcolm Walker, *History of the Meteorological Office* (Cambridge: Cambridge University Press, 2012), xvii.

273 **Given its initial championing:** Richard Schmalensee and Robert N. Stavins, "The S02 Allowance Trading System: The Ironic History of a Grand Policy Experiment," MIT Center for Energy and Environmental Policy Research, August 2012.

273 **"The arrogance of people":** Senator James M. Inhofe, interview, "Crosstalk," Voice of Christian Youth America, March 7, 2013.

274 **The oil and gas industry:** Senator James M. Inhofe, career donations, Center for Responsive Politics, OpenSecrets.Org, federal campaign contributions through March 10, 2014. The exact number is $1,587,596.

274 **Other evangelical Christians:** N. Smith and A. Leiserowitz, "American Evangelicals and Global Warming," *Global Environmental Change,* vol. 23, no. 5 (October 2013), 1009–17.

274 **"It's the poor":** Dr. Katharine Hayhoe, statement to the Rev. Mitch Hescox, "Dr. Katharine Hayhoe Named TIME'S 100 Most Influential People," *Evangelical Environmental Network,* April 24, 2014.

274 **A growing number of scientists:** "Jeff Goodell, Can Dr. Evil Save the World?" *Rolling Stone,* November 2006.

275 **The European Geosciences Union:** "Geoengineering Could Disrupt Rainfall Patterns," European Geosciences Union, June 6, 2012.

275 **As Powell said in his *Arid Lands* report:** Powell, *Report on the Lands of the Arid Region,* 70.

275 **The impossible caveat:** Meyer, *Americans and Their Weather,* 138.

276 **In *The Martian Chronicles*:** Ray Bradbury, *The Martian Chronicles* (New York: Bantam Books, 1979), 67, 68.

EPILOGUE: WAITING FOR RAIN

278   **Between August 1860 and July 1861:** *World Weather Records, Smithsonian Miscellaneous Collections,* vol. 79, "Cherrapunji, India, Precipitation in Inches" (Washington, D.C.: The Smithsonian Institute, 1929), 246.

278   **As the clouds meet the Khasi Hills:** Ahrens, *Essentials of Meteorology,* 350.

279   **But the locals were so hostile:** Alexander Frater, *Chasing the Monsoon: A Modern Pilgrimage Through India* (London: Viking, 1990), 251–61.

281   **Scientists have identified 250 orchid species:** Meghalaya State Development Report, chapter 3, "Development and Management of Natural Resources," 31–40, http://megplanning.gov.in/MSDR/natural_resources.pdf.

287   **Florida is drowning in the rainiest summer:** National Climatic Data Center, NOAA, "National Overview, Summer 2013," http://www.ncdc.noaa .gov/sotc/national/2013/8.

288   **Local governments shuttered the public schools:** Hemanta Kumar Nath, "Assam Heat Wave: Death Toll Rise to 26," India Blooms News Service, June 13, 2013.

291   **By the time I left India:** "India Raises Flood Death Toll, Reaches 5,700 as All Missing Persons Now Presumed Dead," CBS News, July 16, 2013, http://www.cbsnews.com/news/india-raises-flood-death-toll-reaches-5700 -as-all-missing-persons-now-presumed-dead/.

292   **The "drip-tips" guide rainfall:** Nalini Nadkarni, *Between Earth and Sky: Our Intimate Connections to Trees* (Berkeley: University of California Press, 2009), 226.

CYNTHIA BARNETT is an award-winning environmental journalist who has reported on water from the Suwannee River to Singapore. She is the author of *Mirage,* which won the gold medal for best nonfiction in the Florida Book Awards, and *Blue Revolution,* named by the *Boston Globe* as one of the top ten science books of 2011. She lives in Gainesville, Florida, with her husband and their two children.